Industrial Engineering and Manufacturing Processes

Industrial Engineering and Manufacturing Processes

Edited by
Eden Witherspoon

Larsen & Keller
www.larsen-keller.com

Industrial Engineering and Manufacturing Processes
Edited by Eden Witherspoon
ISBN: 978-1-63549-005-3 (Hardback)

■ Larsen & Keller

Published by Larsen and Keller Education,
5 Penn Plaza,
19th Floor,
New York, NY 10001, USA

Cataloging-in-Publication Data

Industrial engineering and manufacturing processes / edited by Eden Witherspoon.
    p. cm.
Includes bibliographical references and index.
ISBN 978-1-63549-005-3
1. Industrial engineering--Textbooks. 2. Production engineering--
Textbooks. 3. Manufacturing processes--Textbooks. I. Witherspoon, Eden.
G155.A1 T68 2017
910.68--dc23

The publisher's policy is to use permanent paper from mills that operate a sustainable forestry policy. Furthermore, the publisher ensures that the text paper and cover boards used have met acceptable environmental accreditation standards.

Printed and bound in the United States of America.

For more information regarding Larsen and Keller Education and its products, please visit the publisher's website www.larsen-keller.com

Table of Contents

Preface

Industrial engineering is primarily concerned with the optimization of all processes involved in an industrial set-up such as material procurement, manufacturing, etc. This book discusses all the theoretical and managerial aspects of industrial engineering. It also analyses the different facets of the end-product like design and development. The methods and processes associated with industrial engineering have also been covered with an overview on different industrial systems. The aim of this textbook is to provide an in-depth knowledge about the field of industrial engineering and its related branches.

Given below is the chapter wise description of the book:

Chapter 1- Industrial engineering is an emerging field of study. The following chapter will not only provide an overview but also delve deep into the varied topics related to it. Along with comprehensive insights to the crucial aspects of Industrial engineering it also lays emphasis on manufacturing, as they are interrelated fields of study.

Chapter 2- Industrial engineering is an interdisciplinary field of study. Hence, to understand it completely it is essential to also study its integrated aspects. The aim of this chapter is to lucidly elaborate topics such as manufacturing engineering, production engineering, process engineering, etc to broaden the scope of knowledge for the readers.

Chapter 3- There are many prevalent theories related to industrial engineering. The most significant out of them is the theory of constraints which has been lucidly covered in this chapter. Another extremely crucial aspect of industrial engineering is planning. This chapter throws light on the planning process involved in industrial engineering, especially in terms of material requirements.

Chapter 4- In order to understand any branch of engineering it is fundamental to understand its varied methods and processes. This chapter will provide an in-depth knowledge about the methods and processes involved in industrial engineering which result in higher efficiency and output. Topics like methods engineering, flow process chart, work measurement, work sampling, etc. are thoroughly discussed in this chapter.

Chapter 5- Research and management are significant areas of study under the branch of industrial engineering. The following chapter unfolds its crucial aspects in a critical yet systematic manner. Major elements like operations research and operations management are discussed in this chapter.

Chapter 6- One of the most important facets of industrial engineering is the product. The following chapter will provide an insightful account on the various processes through which a product takes its final shape. Processes like industrial design, production planning, engineering design process and quality assurance are described thoroughly in the following chapter.

Chapter 7- Industrial engineering is an umbrella discipline which branches out into unique sub-disciplines which can be studied individually as well as in conjunction with industrial engineering. This all-inclusive chapter elucidates the most significant branches of industrial engineering such as safety engineering, business engineering, enterprise engineering, and engineering management amongst others.

Chapter 8- Industrial ecology has emerged as an important area of study in the recent decade. This field also focuses on effective ways to reduce industrial waste. The following chapter will present to the reader all the significant aspects of industrial ecology and also elaborate the history of this field for a better understanding.

Chapter 9- Industries are an amalgamation of systems, in order to understand industries as a whole, it is essential that the readers develop a comprehensive understanding about its different systems. This chapter will focus on the various industrial systems like pilot plant, demonstration plant and industrial robot. The chapter will be an invaluable source for readers as it will broaden their knowledge of this area of study.

Chapter 10- The origin of industrial engineering can be traced back to the industrial revolution. It resulted in the expansion of the industrial sector and also gave it the modern identity which it has today. Thus, it is crucial to have an understanding of the industrial revolution for a better perspective of industrial engineering. This chapter explores some of the most significant aspects of industrial revolution and its impact on industrial engineering.

At the end, I would like to thank all those who dedicated their time and efforts for the successful completion of this book. I also wish to convey my gratitude towards my friends and family who supported me at every step.

Editor

An Introduction to Industrial Engineering and Manufacturing

Industrial engineering is an emerging field of study. The following chapter will not only provide an overview but also delve deep into the varied topics related to it. Along with comprehensive insights to the crucial aspects of Industrial engineering it also lays emphasis on manufacturing, as they are interrelated fields of study.

Industrial Engineering

Industrial engineering is a branch of engineering which deals with the optimization of complex processes, systems or organizations. Industrial engineers work to eliminate waste of time, money, materials, man-hours, machine time, energy and other resources that do not generate value. According to the Institute of Industrial and Systems Engineers, they figure out how to do things better, they engineer processes and systems that improve quality and productivity.

Industrial engineering is concerned with the development, improvement, and implementation of integrated systems of people, money, knowledge, information, equipment, energy, materials, analysis and synthesis, as well as the mathematical, physical and social sciences together with the principles and methods of engineering design to specify, predict, and evaluate the results to be obtained from such systems or processes. While industrial engineering is a longstanding engineering discipline subject to (and eligible for) professional engineering licensure in most jurisdictions, its underlying concepts overlap considerably with certain business-oriented disciplines such as operations management.

Depending on the sub-specialties involved, industrial engineering may also be known as, or overlap with, operations research, systems engineering, manufacturing engineering, management science, management engineering, business engineering, ergonomics or human factors engineering, safety engineering, or others, depending on the viewpoint or motives of the user.

Overview

While originally applied to manufacturing, the use of "industrial" in "industrial engineering" can be somewhat misleading, since it has grown to encompass any methodical or quantitative approach to optimizing how a process, system, or organization operates. Some engineering universities and educational agencies around the world have changed the term "industrial" to broader terms such as "production" or "systems", leading to the typical extensions noted above. In fact, the primary U.S. professional organization for Industrial Engineers, the Institute of Industrial Engineers (IIE) has been considering changing its name to something broader (such as the Institute of Industrial & Systems Engineers), although the latest vote among membership deemed this unnecessary for the time being.

The various topics concerning industrial engineers include:

- accounting: the measurement, processing and communication of financial information about economic entities

- operations research, also known as management science: discipline that deals with the application of advanced analytical methods to help make better decisions

- operations management: an area of management concerned with overseeing, designing, and controlling the process of production and redesigning business operations in the production of goods or services.

- project management: is the process and activity of planning, organizing, motivating, and controlling resources, procedures and protocols to achieve specific goals in scientific or daily problems.

- job design: the specification of contents, methods and relationship of jobs in order to satisfy technological and organizational requirements as well as the social and personal requirements of the job holder.

- financial engineering: the application of technical methods, especially from mathematical finance and computational finance, in the practice of finance

- management engineering: a specialized form of management that is concerned with the application of engineering principles to business practice

- supply chain management: the management of the flow of goods. It includes the movement and storage of raw materials, work-in-process inventory, and finished goods from point of origin to point of consumption.

- process engineering: design, operation, control, and optimization of chemical, physical, and biological processes.

- systems engineering: an interdisciplinary field of engineering that focuses on how to design and manage complex engineering systems over their life cycles.

- ergonomics: the practice of designing products, systems or processes to take proper account of the interaction between them and the people that use them.

- safety engineering: an engineering discipline which assures that engineered systems provide acceptable levels of safety.

- cost engineering: practice devoted to the management of project cost, involving such activities as cost- and control- estimating, which is cost control and cost forecasting, investment appraisal, and risk analysis.

- value engineering: a systematic method to improve the "value" of goods or products and services by using an examination of function.

- quality engineering: a way of preventing mistakes or defects in manufactured products and avoiding problems when delivering solutions or services to customers.

- Industrial plant configuration: sizing of necessary infrastructure used in support and maintenance of a given facility.

- facility management: an interdisciplinary field devoted to the coordination of space, infrastructure, people and organization

- engineering design process: formulation of a plan to help an engineer build a product with a specified performance goal.

- logistics: the management of the flow of goods between the point of origin and the point of consumption in order to meet some requirements, of customers or corporations.

Traditionally, a major aspect of industrial engineering was planning the layouts of factories and designing assembly lines and other manufacturing paradigms. And now, in so-called lean manufacturing systems, industrial engineers work to eliminate wastes of time, money, materials, energy, and other resources.

Examples of where industrial engineering might be used include flow process charting, process mapping, designing an assembly workstation, strategizing for various operational logistics, consulting as an efficiency expert, developing a new financial algorithm or loan system for a bank, streamlining operation and emergency room location or usage in a hospital, planning complex distribution schemes for materials or products (referred to as supply-chain management), and shortening lines (or queues) at a bank, hospital, or a theme park.

Modern industrial engineers typically use predetermined motion time system, computer simulation (especially discrete event simulation), along with extensive mathematical tools for modelling, such as mathematical optimization and queue theory, and computational methods for system analysis, evaluation, and optimization.

History

Industrial Revolution

There is a general consensus among historian that the roots of the Industrial Engineering Profession date back to the Industrial Revolution. The technologies that helped mechanize traditional manual operations in the textile industry including the Flying shuttle, the Spinning jenny, and perhaps most importantly the Steam engine generated Economies of scale that made Mass production of in centralized locations attractive for the first time. The concept of the production system had its genesis in the factories created by these innovations.

Specialization of Labor

Adam Smith's concepts of Division of Labour and the "Invisible Hand" of capitalism introduced in his treatise "The Wealth of Nations" motivated many of the technological innovators of the Industrial revolution to establish and implement factory systems. The efforts of James Watt and Matthew Boulton led to the first integrated machine manufacturing facility in the world, including the implementation of concepts such as cost control systems to reduce waste and increase productivity and the institution of skills training for craftsmen.

Charles Babbage became associated with Industrial engineering because of the concepts he introduced in his book "On the Economy of Machinery and Manufacturers" which he wrote as a result of his visits to factories in England and the United States in the early 1800s. The book includes subjects such as the time required to perform a specific task, the effects of subdividing tasks into smaller and less detailed elements, and the advantages to be gained from repetitive tasks.

Interchangeable Parts

Eli Whitney and Simeon North proved the feasibility of the notion of Interchangeable parts in the manufacture of muskets and pistols for the US Government. Under this system, individual parts were mass-produced to tolerances to enable their use in any finished product. The result was a significant reduction in the need for skill from specialized workers, which eventually led to the industrial environment to be studied later.

Pioneers

Frederick Tayl`or is generally credited as being the father of the Industrial Engineering discipline. He earned a degree in mechanical engineering from Steven's University, and earned several patents from his inventions. His books, *Shop Management* and *The Principles of Scientific management* which were published in the early 1900s, were the beginning of Industrial Engineering. Improvements in work efficiency under his methods was based on improving work methods, developing of work standards, and reduction in time required to carry out the work. With an abiding faith in the scientific method, Taylor's contribution to "Time Study" sought a high level of precision and predictability for manual tasks.

Frank Gilbreth and Lilian Gilbreth were the other corner stone of the Industrial Engineering movement. They categorized the elements of human motion into 18 basic elements called *therbligs*. This development permitted analysts to design jobs without knowledge of the time required to do a job. These developments were the beginning of a much broader field known as human factors or ergonomics.

In the United States, the first department of industrial and manufacturing engineering was established at the Pennsylvania State University in 1909. The first doctoral degree in industrial engineering was awarded in 1933 by Cornell University.

In 1912 Henry Laurence Gantt developed the Gantt chart which outlines actions the organization along with their relationships. This chart opens later form familiar to us today by Wallace Clark.

Assembly lines: moving car factory of Henry Ford (1913) accounted for a significant leap forward in the field. Ford reduced the assembly time of a car more than 700 hours to 1.5 hours. In addition, he was a pioneer of the economy of the capitalist welfare ("welfare capitalism") and the flag of providing financial incentives for employees to increase productivity.

Comprehensive quality management system (Total quality management or TQM) developed in the forties was gaining momentum after World War II and was part of the recovery of Japan after the war.

Modern Practice

In 1960 to 1975, with the development of decision support systems in supply such as the Material requirements planning (MRP), you can emphasize the timing issue (inventory, production, compounding, transportation, etc.) of industrial organization. Israeli scientist Dr. Jacob Rubinovitz installed the CMMS program developed in IAI and Control-Data (Israel) in 1976 in South Africa and worldwide.

In the seventies, with the penetration of Japanese management theories such as Kaizen and Kanban, Japan realized very high levels of quality and productivity. These theories improved issues of quality, delivery time, and flexibility. Companies in the west realized the great impact of Kaizen and started implementing their own Continuous improvement programs.

In the nineties, following the global industry globalization process, the emphasis was on supply chain management, and customer-oriented business process design. Theory of constraints developed by an Israeli scientist Eliyahu M. Goldratt (1985) is also a significant milestone in the field.

Compared to Other Engineering Disciplines

Engineering is traditionally decompositional. To understand the whole, it is first broken into its parts. One then masters the parts and puts them back together, becoming the master of the whole. Industrial and systems engineering's (ISE) approach is the opposite; any one part cannot be understood without the context of the whole. Changes in one part affect the whole, and the role of a part is a projection into the whole. In traditional engineering, people understand the parts first, then they can understand the whole. In ISE, they understand the whole first, and then they can understand the role of each part.

University Programs

2016 U.S. News Undergraduate Rankings	
University	**Rank**
Georgia Institute of Technology	1
University of Michigan, Ann Arbor	2
Purdue University	3
University of California, Berkeley	3
Northwestern University	5
Virginia Tech	6
Penn State University	7
Stanford University	8
University of Wisconsin-Madison	9

Universities offer degrees at the bachelor, masters, and doctoral level.

Undergraduate Curriculum

In the United States the undergraduate degree earned is the Bachelor of Science (B.S.) or Bachelor of Science and Engineering (B.S.E.) in Industrial Engineering (IE). Variations of the title include Industrial & Operations Engineering (IOE), and Industrial & Systems Engineering (ISE). The typical curriculum includes a broad math and science foundation spanning chemistry, physics, mechanics (i.e. statics and dynamics), materials science, computer science, electronics/circuits, engineering design, and the standard range of engineering mathematics (i.e. calculus, differential equations, statistics). For any engineering undergraduate program to be accredited, regardless of concentration, it must cover a largely similar span of such foundational work - which also overlaps heavily with the content tested on one or more engineering licensure exams in most jurisdictions.

The coursework specific to IE entails specialized courses in areas such as systems theory, Ergonomics/safety, Stochastic modeling, optimization, and engineering economics. Elective subjects may include management, finance, strategy, and other business-oriented courses, and for general electives social science and humanities courses. Business schools may offer programs with some overlapping relevance to IE, but the engineering programs are distinguished by a more intensely quantitative focus as well as the core math and science courses required of all engineering programs.

Postgraduate Curriculum

The usual postgraduate degree earned is the Master of Science (MS) or Master of Science and Engineering (MSE) in Industrial Engineering or various alternative related concentration titles. Typical MS curricula may cover:

• Operations research and optimization techniques	• Facilities design and work-space design
• Engineering economics	• Quality engineering
• Supply chain management and logistics	• Reliability engineering and life testing
• Systems simulation and stochastic processes	• Statistical process control or quality control
• System dynamics and policy planning	• Time and motion study
• System analysis and techniques	• Predetermined motion time system and computer use for IE
• Manufacturing systems/manufacturing engineering	• Operations management
• Human factors engineering and ergonomics (safety engineering)	• Corporate planning
• Production planning and control	• Productivity improvement
• Management sciences	• Materials management
• Computer-aided manufacturing	• Robotics
• Lean Six Sigma	• Product Development

Salaries and Workforce Statistics

United States

The total number of engineers employed in the US in 2006 was roughly 1.5 million. Of these, 201,000 were industrial engineers (13.3%), the third most popular engineering specialty. The average starting salaries were $55,067 with a bachelor's degree, $77,364 with a master's degree, and $100,759 with a doctorate degree. This places industrial engineering at 7th of 15 among engineering bachelor's degrees, 3rd of 10 among master's degrees, and 2nd of 7 among doctorate degrees in average annual salary. The median annual income of industrial engineers in the U.S. workforce is $68,624.

Norway

The average total starting salary in 2011 for Norwegian industrial engineers is NOK 505,100 ($83,100), while the average total salary in general is NOK 1,049,054 ($172 600).

Related topics	Associations
• Operations research	• Institute of Industrial and Systems Engineers
• Systems engineering	• INFORMS
• Engineering management	• Institute of Industrial and Systems Engineers
• Manufacturing engineering	• American Society for Engineering Education
• Operations engineering	• American Society for Quality
• Enterprise engineering	• The Australian Society for Operations Research
• Maintenance engineering	• The UK MTM Association
• Production engineering	• European Students of Industrial Engineering and Management
• Quality engineering	• The International Federation of Operational Research Societies (IFORS)
• Human factors engineering	• Indian Institution of Industrial Engineering
• Project management	• Iranian Institute of Industrial Engineering
• Safety engineering	• Washington Accord
• Engineering economics	• The Operations Research Society of Japan
• Environment, health and safety	
• List of production topics	
• Nutrient systems	
• Overall equipment effectiveness	
• Product design / industrial design	
• Reverse engineering	
• Occupational safety and health	
• Sales process engineering	
• Sociotechnical systems	
• Statistical process control	
• Toyota production system	

Industrial and Production Engineering

Industrial and production engineering is a combination of mechanical engineering and industrial engineering and management science. It is a branch of engineering concerned with the development, improvement, implementation and evaluation of integrated systems of people, money, knowledge, information, equipment, energy, material and process. It is very necessary for any manufacturing company and service providing enterprise to implement this concept of engineering and principles of management science. there are many universities all over the world provide Bachelor and Master and Phd degrees in this engineering course. In Europe,America and other part of worlds Industrial Engineering and Production Engineering are taught separately.But in Asia there are many universities that provide combined course.

History of Industrial and Production Engineering

This dates back to 1700s. The onset is not very clear but the ideology started developing the pioneers of this field who laid the foundation were Sir Adam Smith, Henry Ford, Eli Whitney, Frank and Lilian Gilbreth, Henry Gantt, F.W. Taylor, etc. Industrial and production engineering were practiced as two separate entities. Much later the world thought of combining this two fields, hence came the branch of industrial and production engineering though different countries have named the branch differently but the content remains unaltered its almost same. In developed countries it's known as industrial engineering and in south Asian countries it is popularly known as industrial and production engineering (India, Bangladesh).

History of Industrial and Production Engineering in South Asia

In developed countries like the US this subject was initiated long ago. In 1974 the Indian institute of technology started an undergraduate programme in industrial engineering But in third world countries like Bangladesh this subject was not known. It was started in 1981 as a specialized part of mechanical engineering. Only M.Sc. was offered then. In 1997 it was made fully independent and separated department as B.Sc. engineering course was initiated.

Overview

This field also deals with the integration of different facilities and systems for producing quality products (with optimal expenditure) by applying the principles of physics and the results of manufacturing systems studies, such as the following:

• Craft or guild	• Computer integrated manufacturing	• Agile manufacturing
• Putting-out system	• Computer-aided technologies in manufacturing	• Rapid manufacturing
• English system of manufacturing	• Just in time manufacturing	• Prefabrication
• American system of manufacturing	• Lean manufacturing	• Ownership
• Soviet collectivism in manufacturing	• Flexible manufacturing	• Fabrication
• Mass production	• Mass customization	• Publication

Manufacturing engineers develop and create physical artifacts, production processes, and technology. It is a very broad area which includes the design and development of products. The manufacturing engineering discipline has very strong overlaps with mechanical engineering, industrial engineering, production engineering, electrical engineering, electronic engineering, computer science, materials management, and operations management. Manufacturing engineers' success or failure directly impacts the advancement of technology and the spread of innovation. This field of manufacturing engineering emerged from tool and die discipline in the early 20th century. It expanded greatly from the 1960s when industrialized countries introduced factories with:

1. Numerical control machine tools and automated systems of production.

2. Advanced statistical methods of quality control: These factories were pioneered by the American electrical engineer William Edwards Deming, who was initially ignored by his home country. The same methods of quality control later turned Japanese factories into world leaders in cost-effectiveness and production quality.

3. Industrial robots on the factory floor, introduced in the late 1970s: These computer-controlled welding arms and grippers could perform simple tasks such as attaching a car door quickly and flawlessly 24 hours a day. This cut costs and improved production speed.

History of Manufacturing Engineering

The history of manufacturing engineering can be traced to factories in the mid 19th century USA and 18th century UK. Although large home production sites and workshops were established in ancient China, ancient Rome and the Middle East, the Venice Arsenal provides one of the first examples of a factory in the modern sense of the word. Founded in 1104 in the Republic of Venice several hundred years before the Industrial Revolution, this factory mass-produced ships on assembly lines using manufactured parts. The Venice Arsenal apparently produced nearly one ship every day and, at its height, employed 16,000 people.

Many historians regard Matthew Boulton's Soho Manufactory (established in 1761 in Birmingham) as the first modern factory. Similar claims can be made for John Lombe's silk mill in Derby (1721), or Richard Arkwright's Cromford Mill (1771). The Cromford Mill was purpose-built to accommodate the equipment it held and to take the material through the various manufacturing processes.

Ford assembly line, 1913.

One historian, Murno Gladst, contends that the first factory was in Potosí. The Potosi factory took advantage of the abundant silver that was mined nearby and processed silver ingot slugs into coins.

British colonies in the 19th century built factories simply as buildings where a large number of workers gathered to perform hand labor, usually in textile production. This proved more efficient for the administration and distribution of materials to individual workers than earlier methods of manufacturing, such as cottage industries or the putting-out system.

Cotton mills used inventions such as the steam engine and the power loom to pioneer the industrial factories of the 19th century, where precision machine tools and replaceable parts allowed greater efficiency and less waste. This experience formed the basis for the later studies of manufacturing engineering. Between 1820 and 1850, non-mechanized factories supplanted traditional artisan shops as the predominant form of manufacturing institution.

Henry Ford further revolutionized the factory concept and thus manufacturing engineering in the early 20th century with the innovation of mass production. Highly specialized workers situated alongside a series of rolling ramps would build up a product such as (in Ford's case) an automobile. This concept dramatically decreased production costs for virtually all manufactured goods and brought about the age of consumerism.

Modern Developments

Modern manufacturing engineering studies include all intermediate processes required for the production and integration of a product's components.

Some industries, such as semiconductor and steel manufacturers use the term "fabrication" for these processes.

KUKA industrial robots being used at a bakery for food production

Automation is used in different processes of manufacturing such as machining and welding. Automated manufacturing refers to the application of automation to produce goods in a factory. The main advantages of automated manufacturing for the manufacturing process are realized with effective implementation of automation and include: higher consistency and quality, reduction of lead times, simplification of production, reduced handling, improved work flow, and improved worker morale.

Robotics is the application of mechatronics and automation to create robots, which are often used in manufacturing to perform tasks that are dangerous, unpleasant, or repetitive. These robots may be of any shape and size, but all are pre-programmed and interact physically with the world. To create a robot, an engineer typically employs kinematics (to determine the robot's range of motion) and mechanics (to determine the stresses within the robot). Robots are used extensively in manufacturing engineering.

Robots allow businesses to save money on labor, perform tasks that are either too dangerous or too precise for humans to perform economically, and to ensure better quality. Many companies employ assembly lines of robots, and some factories are so robotized that they can run by themselves. Outside the factory, robots have been employed in bomb disposal, space exploration, and many other fields. Robots are also sold for various residential applications.

Education

Certification Programs in Manufacturing Engineering

Manufacturing engineers possess a bachelor's degree in engineering with a major in manufacturing engineering. The length of study for such a degree is usually four to five years followed by five more years of professional practice to qualify as a professional engineer. Working as a manufacturing engineering technologist involves a more applications-oriented qualification path.

Academic degrees for manufacturing engineers are usually the Bachelor of Engineering, [BE] or [BEng], and the Bachelor of Science, [BS] or [BSc]. For manufacturing technologists the required degrees are Bachelor of Technology [B.TECH] or Bachelor of Applied Science [BASc] in Manufacturing, depending upon the university. Masters degrees in engineering manufacturing include Master of Engineering [ME] or [MEng] in Manufacturing, Master of Science [M.Sc] in Manufacturing Management, Master of Science [M.Sc] in Industrial and Production Management, and Master of Science [M.Sc] as well as Master of Engineering [ME] in Design, which is a subdiscipline of manufacturing. Doctoral [PhD] or [DEng] level courses in manufacturing are also available depending on the university.

The undergraduate degree curriculum generally includes courses in physics, mathematics, computer science, project management, and specific topics in mechanical and manufacturing engineering. Initially such topics cover most, if not all, of the subdisciplines of manufacturing engineering. Students then choose to specialize in one or more subdisciplines towards the end of their degree work.

Syllabus

The foundational curriculum for a bachelor's degree in manufacturing engineering and includes: Environment, health and safety

- Human factors engineering
- Living wage
- List of production topics
- Nutrient systems

- Overall equipment effectiveness
- Product design / industrial design
- Production engineering
- Reverse engineering
- Occupational safety and health
- Sales process engineering
- Statistical process control
- Operation Management
- Quality control and management
- Computer-integrated manufacturing
- CAD/CAM
- Supply chain management
- Statics and dynamics
- Strength of materials and solid mechanics
- Instrumentation and measurement
- Applied thermodynamics, heat transfer, energy conversion, and HVAC
- Fluid mechanics and fluid dynamics
- Mechanism design (including kinematics and dynamics)
- Manufacturing technology or processes
- Hydraulics and pneumatics
- Mathematics – in particular, calculus, differential equations, statistics, and linear algebra.
- Engineering design and graphics
- Circuit Analysis
- Lean manufacturing
- Mechatronics and control theory
- Automation and reverse engineering
- Quality assurance and control
- Material science
- Drafting, CAD (including solid modeling), and CAM, etc.

A bachelor's degree in these two areas will typically differ only by a few specialized classes, although the mechanical engineering degree requires more mathematics expertise.

Manufacturing Engineering Certification

Certification and Licensure:

In some countries, "professional engineer" is the term for registered or licensed engineers who are permitted to offer their professional services directly to the public. Professional Engineer, abbreviatied (PE - USA) or (PEng - Canada), is the designation for licensure in North America. In order to qualify for this license, a candidate needs a bachelor's degree from an ABET recognized university in the USA, a passing score on a state examination, and four years of work experience usually gained via a structured internship. In the USA, more recent graduates have the option of dividing this licensure process into two segments. The Fundamentals of Engineering (FE) exam is often taken immediately after graduation and the Principles and Practice of Engineering exam is taken after four years of working in a chosen engineering field.

Society of Manufacturing Engineers (SME) certifications (USA):

The SME administers qualifications specifically for the manufacturing industry. These are not degree level qualifications and are not recognized at the professional engineering level. The following discussion deals with qualifications in the US only. Qualified candidates for the Certified Manufacturing Technologist Certificate (CMfgT) must pass a three-hour, 130-question multiple-choice exam. The exam covers maththematics, manufacturing processes, manufacturing management, automation, and related subjects. Additionally, a candidate must have at least four years of combined education and manufacturing-related work experience.

Certified Manufacturing Engineer (CMfgE) is an engineering qualification administered by the Society of Manufacturing Engineers, Dearborn, Michigan, US. Candidates qualifying for a Certified Manufacturing Engineer credential must pass a four-hour, 180 question multiple-choice exam which covers more in-depth topics than does the CMfgT exam. CMfgE candidates must also have eight years of combined education and manufacturing-related work experience, with a minimum of four years of work experience.

Certified Engineering Manager (CEM). The Certified Engineering Manager Certificate is also designed for engineers with eight years of combined education and manufacturing experience. The test is four hours long and has 160 multiple-choice questions. The CEM certification exam covers business processes, teamwork, responsibility, and other management-related categories.

Modern Tools

Many manufacturing companies, especially those in industrialized nations, have begun to incorporate computer-aided engineering (CAE) programs into their existing design and analysis processes, including 2D and 3D solid modeling computer-aided design (CAD). This method has many benefits, including easier and more exhaustive visualization of products, the ability to create virtual assemblies of parts, and ease of use in designing mating interfaces and tolerances.

Other CAE programs commonly used by product manufacturers include product life cycle management (PLM) tools and analysis tools used to perform complex simulations. Analysis tools may be used to predict product response to expected loads, including fatigue life and manufacturability. These tools include finite element analysis (FEA), computational fluid dynamics (CFD), and computer-aided manufacturing (CAM).

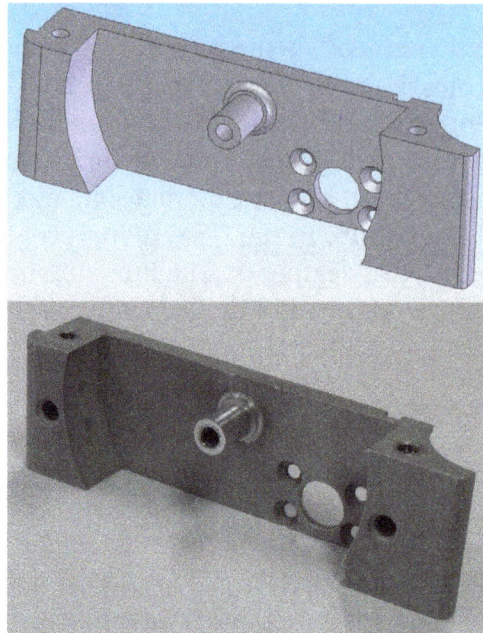

CAD model and CNC machined part

Using CAE programs, a mechanical design team can quickly and cheaply iterate the design process to develop a product that better meets cost, performance, and other constraints. No physical prototype need be created until the design nears completion, allowing hundreds or thousands of designs to be evaluated, instead of relatively few. In addition, CAE analysis programs can model complicated physical phenomena which cannot be solved by hand, such as viscoelasticity, complex contact between mating parts, or non-Newtonian flows.

Just as manufacturing engineering is linked with other disciplines, such as mechatronics, multidisciplinary design optimization (MDO) is also being used with other CAE programs to automate and improve the iterative design process. MDO tools wrap around existing CAE processes, allowing product evaluation to continue even after the analyst goes home for the day. They also utilize sophisticated optimization algorithms to more intelligently explore possible designs, often finding better, innovative solutions to difficult multidisciplinary design problems.

Subdisciplines

Mechanics

Mechanics, in the most general sense, is the study of forces and their effects on matter. Typically, engineering mechanics is used to analyze and predict the acceleration and deformation (both elastic and plastic) of objects under known forces (also called loads) or stresses. Subdisciplines of mechanics include:

- Statics, the study of non-moving bodies under known loads

- Dynamics (or kinetics), the study of how forces affect moving bodies

- Mechanics of materials, the study of how different materials deform under various types of stress

- Fluid mechanics, the study of how fluids react to forces

- Continuum mechanics, a method of applying mechanics that assumes that objects are continuous (rather than discrete)

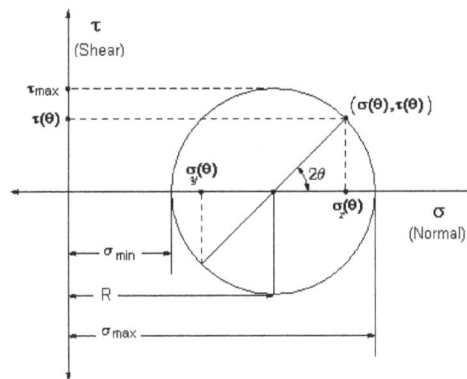

Mohr's circle, a common tool to study stresses in a mechanical element

If the engineering project were to design a vehicle, statics might be employed to design the frame of the vehicle in order to evaluate where the stresses will be most intense. Dynamics might be used when designing the car's engine to evaluate the forces in the pistons and cams as the engine cycles. Mechanics of materials might be used to choose appropriate materials for the manufacture of the frame and engine. Fluid mechanics might be used to design a ventilation system for the vehicle or to design the intake system for the engine.

Kinematics

Kinematics is the study of the motion of bodies (objects) and systems (groups of objects), while ignoring the forces that cause the motion. The movement of a crane and the oscillations of a piston in an engine are both simple kinematic systems. The crane is a type of open kinematic chain, while the piston is part of a closed four-bar linkage. Engineers typically use kinematics in the design and analysis of mechanisms. Kinematics can be used to find the possible range of motion for a given mechanism, or, working in reverse, can be used to design a mechanism that has a desired range of motion.

Drafting

Drafting or technical drawing is the means by which manufacturers create instructions for manufacturing parts. A technical drawing can be a computer model or hand-drawn schematic showing all the dimensions necessary to manufacture a part, as well as assembly notes, a list of required materials, and other pertinent information. A U.S engineer or skilled worker who creates technical drawings may be referred to as a drafter or draftsman. Drafting has historically been a two-dimensional process, but computer-aided design (CAD) programs now allow the designer to create in three dimensions.

Instructions for manufacturing a part must be fed to the necessary machinery, either manually, through programmed instructions, or through the use of a computer-aided manufacturing (CAM) or combined CAD/CAM program. Optionally, an engineer may also manually manufacture a part using the technical drawings, but this is becoming an increasing rarity with the advent of computer numerically controlled (CNC) manufacturing. Engineers primarily manufacture parts manually in the areas of applied spray coatings, finishes, and other processes that cannot economically or practically be done by a machine.

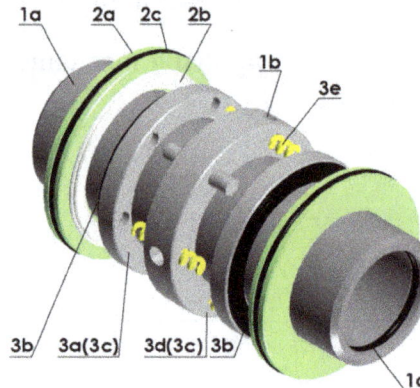

A CAD model of a mechanical double seal

Drafting is used in nearly every subdiscipline of mechanical and manufacturing engineering, and by many other branches of engineering and architecture. Three-dimensional models created using CAD software are also commonly used in finite element analysis (FEA) and computational fluid dynamics (CFD).

Mechatronics

Training FMS with learning robot SCORBOT-ER 4u, workbench CNC mill and CNC lathe

Mechatronics is an engineering discipline that deals with the convergence of electrical, mechanical and manufacturing systems. Such combined systems are known as electromechanical systems and are widespread. Examples include automated manufacturing systems, heating, ventilation and air-conditioning systems, and various aircraft and automobile subsystems.

The term mechatronics is typically used to refer to macroscopic systems, but futurists have predicted the emergence of very small electromechanical devices. Already such small devices, known as Microelectromechanical systems (MEMS), are used in automobiles to initiate the deployment of airbags, in digital projectors to create sharper images, and in inkjet printers to create nozzles for high-definition printing. In the future it is hoped that such devices will be used in tiny implantable medical devices and to improve optical communication.

Textile Engineering

Textile engineering courses deal with the application of scientific and engineering principles to the design and control of all aspects of fiber, textile, and apparel processes, products, and machinery. These include natural and man-made materials, interaction of materials with machines, safety and health, energy conservation, and waste and pollution control. Additionally, students are given experience in plant design and layout, machine and wet process design and improvement, and designing and creating textile products. Throughout the textile engineering curriculum, students take classes from other engineering and disciplines including: mechanical, chemical, materials and industrial engineering.

Employment

Manufacturing engineering is just one facet of the engineering industry. Manufacturing engineers enjoy improving the production process from start to finish. They have the ability to keep the whole production process in mind as they focus on a particular portion of the process. Successful students in manufacturing engineering degree programs are inspired by the notion of starting with a natural resource, such as a block of wood, and ending with a usable, valuable product, such as a desk, produced efficiently and economically.

Manufacturing engineers are closely connected with engineering and industrial design efforts. Examples of major companies that employ manufacturing engineers in the United States include General Motors Corporation, Ford Motor Company, Chrysler, Boeing, Gates Corporation and Pfizer. Examples in Europe include Airbus, Daimler, BMW, Fiat, Navistar International, and Michelin Tyre.

Industries where manufacturing engineers are generally employed include:

- Aerospace industry
- Automotive industry
- Chemical industry
- Computer industry
- Electronics industry
- Food processing industry
- Garment industry
- Pharmaceutical industry
- Pulp and paper industry
- Toy industry

Frontiers of Research

Flexible Manufacturing Systems

A typical FMS system

A flexible manufacturing system (FMS) is a manufacturing system in which there is some amount of flexibility that allows the system to react to changes, whether predicted or unpredicted. This flexibility is generally considered to fall into two categories, both of which have numerous sub-categories. The first category, machine flexibility, covers the system's ability to be changed to produce new product types and the ability to change the order of operations executed on a part. The second category, called routing flexibility, consists of the ability to use multiple machines to perform the same operation on a part, as well as the system's ability to absorb large-scale changes, such as in volume, capacity, or capability.

Most FMS systems comprise three main systems. The work machines, which are often automated CNC machines, are connected by a material handling system to optimize parts flow, and to a central control computer, which controls material movements and machine flow. The main advantages of an FMS is its high flexibility in managing manufacturing resources like time and effort in order to manufacture a new product. The best application of an FMS is found in the production of small sets of products from a mass production.

Computer Integrated Manufacturing

Computer-integrated manufacturing (CIM) in engineering is a method of manufacturing in which the entire production process is controlled by computer. Traditionally separated process methods are joined through a computer by CIM. This integration allows the processes to exchange information and to initiate actions. Through this integration, manufacturing can be faster and less error-prone, although the main advantage is the ability to create automated manufacturing processes. Typically CIM relies on closed-loop control processes based on real-time input from sensors. It is also known as flexible design and manufacturing.

Friction Stir Welding

Close-up view of a friction stir weld tack tool

Friction stir welding was discovered in 1991 by The Welding Institute (TWI). This innovative steady state (non-fusion) welding technique joins previously un-weldable materials, including several aluminum alloys. It may play an important role in the future construction of airplanes, potentially replacing rivets. Current uses of this technology to date include: welding the seams of the aluminum main space shuttle external tank, the Orion Crew Vehicle test article, Boeing Delta II and Delta IV Expendable Launch Vehicles and the SpaceX Falcon 1 rocket; armor plating for amphibious assault ships; and welding the wings and fuselage panels of the new Eclipse 500 aircraft from Eclipse Aviation, among an increasingly growing range of uses.

Other areas of research are Product Design, MEMS (Micro-Electro-Mechanical Systems), Lean Manufacturing, Intelligent Manufacturing Systems, Green Manufacturing, Precision Engineering, Smart Materials, etc.

Manufacturing

Manufacturing is the value added production of merchandise for use or sale using labour and machines, tools, chemical and biological processing, or formulation. The term may refer to a range of human activity, from handicraft to high tech, but is most commonly applied to industrial production, in which raw materials are transformed into finished goods on a large scale. Such finished goods may be sold to other manufacturers for the production of other, more complex products, such as aircraft, household appliances or automobiles, or sold to wholesalers, who in turn sell them to retailers, who then sell them to end users and consumers.

Manufacturing or manufacturing process are the steps through which raw materials are transformed into a final product. The manufacturing process begins with the creation of the materials from which the design is made. These materials are then modified through manufacturing processes to become the required part.

Manufacturing takes turns under all types of economic systems. In a free market economy, manufacturing is usually directed toward the mass production of products for sale to consumers at a profit. In a collectivist economy, manufacturing is more frequently directed by the state to supply a centrally planned economy. In mixed market economies, manufacturing occurs under some degree of government regulation.

Modern manufacturing includes all intermediate processes required for the production and integration of a product's components. Some industries, such as semiconductor and steel manufacturers use the term *fabrication* instead.

The manufacturing sector is closely connected with engineering and industrial design. Examples of major manufacturers in North America include General Motors Corporation, General Electric, Procter & Gamble, General Dynamics, Boeing, Pfizer, and Precision Castparts. Examples in Europe include Volkswagen Group, Siemens, and Michelin. Examples in Asia include Sony, Huawei, Lenovo, Toyota, Samsung, and Bridgestone.

History and Development

Finished regenerative thermal oxidizer at manufacturing plant

Assembly of Section 41 of a Boeing 787 Dreamliner

An industrial worker amidst heavy steel semi-products (KINEX BEARINGS, Bytča, Slovakia, c. 1995–2000)

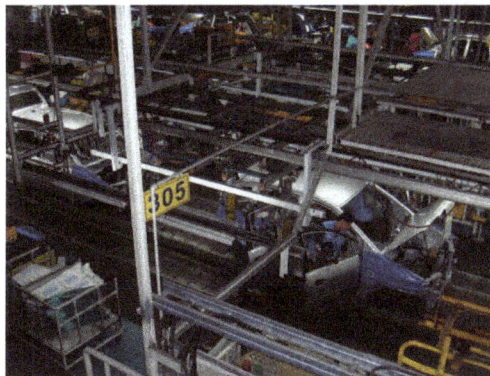

A modern automobile assembly line

- In its earliest form, manufacturing was usually carried out by a single skilled artisan with assistants. Training was by apprenticeship. In much of the pre-industrial world, the guild system protected the privileges and trade secrets of urban artisans.

- Before the Industrial Revolution, most manufacturing occurred in rural areas, where household-based manufacturing served as a supplemental subsistence strategy to agriculture (and continues to do so in places). Entrepreneurs organized a number of manufacturing households into a single enterprise through the putting-out system.

- Toll manufacturing is an arrangement whereby a first firm with specialized equipment processes raw materials or semi-finished goods for a second firm.

Manufacturing systems: changes in methods of manufacturing

- Agile manufacturing

- American system of manufacturing

- British factory system of manufacturing

- Craft or guild system

- Fabrication

- Flexible manufacturing

- Just-in-time manufacturing

- Lean manufacturing

- Mass customization (2000s) - 3D printing, design-your-own web sites for sneakers, fast fashion

- Mass production

- Ownership

- Packaging and labeling

- Prefabrication

- Putting-out system

- Rapid manufacturing

- Reconfigurable manufacturing system

- Soviet collectivism in manufacturing

Industrial Policy

Economics of Manufacturing

According to some economists, manufacturing is a wealth-producing sector of a country, whereas a service sector tends to be wealth-consuming. Emerging technologies have provided some new growth in advanced manufacturing employment opportunities in the Manufacturing Belt in the United States. Manufacturing provides important material support for national infrastructure and for national defense.

On the other hand, most manufacturing may involve significant social and environmental costs. The clean-up costs of hazardous waste, for example, may outweigh the benefits of a product that creates it. Hazardous materials may expose workers to health risks. These costs are now well known and there is effort to address them by improving efficiency, reducing waste, using industrial symbiosis, and eliminating harmful chemicals. The increased use of technologies such as 3D printing also offer the potential to reduce the environmental impact of producing finished goods through distributed manufacturing.

The negative costs of manufacturing can also be addressed legally. Developed countries regulate manufacturing activity with labor laws and environmental laws. Across the globe, manufacturers can be subject to regulations and pollution taxes to offset the environmental costs of manufacturing activities. Labor unions and craft guilds have played a historic role in the negotiation of worker rights and wages. Environment laws and labor protections that are available in developed nations may not be available in the third world. Tort law and product liability impose additional costs on manufacturing. These are significant dynamics in the ongoing process, occurring over the last few decades, of manufacture-based industries relocating operations to "developing-world" economies where the costs of production are significantly lower than in "developed-world" economies.

Manufacturing and Investment

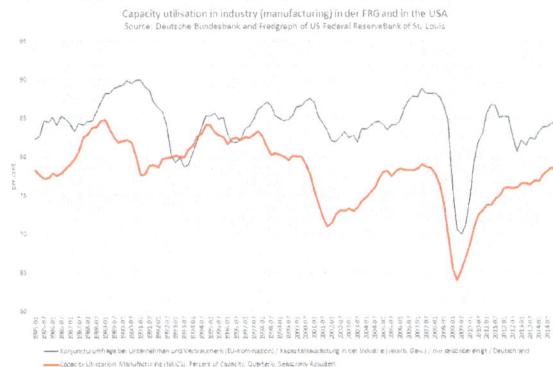

Capacity utilization in manufacturing in the FRG and in the USA

Surveys and analyses of trends and issues in manufacturing and investment around the world focus on such things as:

- The nature and sources of the considerable variations that occur cross-nationally in levels of manufacturing and wider industrial-economic growth;

- Competitiveness; and

- Attractiveness to foreign direct investors.

In addition to general overviews, researchers have examined the features and factors affecting particular key aspects of manufacturing development. They have compared production and investment in a range of Western and non-Western countries and presented case studies of growth and performance in important individual industries and market-economic sectors.

On June 26, 2009, Jeff Immelt, the CEO of General Electric, called for the United States to increase its manufacturing base employment to 20% of the workforce, commenting that the U.S. has outsourced too much in some areas and can no longer rely on the financial sector and consumer spending to drive demand. Further, while U.S. manufacturing performs well compared to the rest of the U.S. economy, research shows that it performs poorly compared to manufacturing in other high-wage countries. A total of 3.2 million – one in six U.S. manufacturing jobs – have disappeared between 2000 and 2007. In the UK, EEF the manufacturers organisation has led calls for the UK economy to be rebalanced to rely less on financial services and has actively promoted the manufacturing agenda.

Countries by Manufacturing Output Using the Most Recent Known Data

List of top 20 manufacturing countries by total value of manufacturing in US dollars for its noted year according to Worldbank.

Rank	Country/Region	Millions of $US	Year
	World	**12,578,627**	**2014**
1	China	3,713,300	2014
9999999	*European Union*	700625660700000000000♣2,566,070	2014
2	United States	2,068,080	2014
9999999	*Eurozone*	700619468570000000000♠1,946,857	2014
3	Japan	850,902	2014
4	Germany	787,503	2014
5	South Korea	389,582	2014
6	India	321,721	2014
7	Italy	296,611	2014
8	France	283,664	2014
9	United Kingdom	282,675	2014
10	Russia	248,481	2014
11	Brazil	218,799	2014
12	Mexico	216,773	2014
13	Indonesia	186,744	2014
14	Spain	166,594	2014
15	Canada	162,074	2014
16	Switzerland	128,881	2014
17	Turkey	126,365	2014
18	Thailand	112,214	2014
19	Netherlands	95,683	2014
20	Australia	93,461	2016

Manufacturing Processes

- List of manufacturing processes
- Manufacturing Process Management

Theories

- Taylorism/Scientific management
- Fordism

Control

- Management
    - o   List of management topics
    - o   Total Quality Management
- Quality control
    - o   Six Sigma

Outline of manufacturing

- List of largest manufacturing companies by revenue
- Industrial robot
- Manufacturing engineering
- Industrial engineering
- Advanced manufacturing
- Metal fabrication
- Microfabrication
- Optics fabrication
- Semiconductor device fabrication
- Biomanufacturing
- Mesoscale Manufacturing
- Cyber manufacturing

References

- *Manufacturing & Investment Around The World: An International Survey Of Factors Affecting Growth & Performance*, ISR Publications/Google Books, revised second edition, 2002. ISBN 978-0-906321-25-6.
- *Research, Industrial Systems (2002-05-20). "Manufacturing and Investment Around the World: An International Survey of Factors Affecting Growth and Performance". ISBN 978-0-906321-25-6.*

Significant Aspects of Industrial Engineering

Industrial engineering is an interdisciplinary field of study. Hence, to understand it completely it is essential to also study its integrated aspects. The aim of this chapter is to lucidly elaborate topics such as manufacturing engineering, production engineering, process engineering, etc to broaden the scope of knowledge for the readers.

Systems Engineering

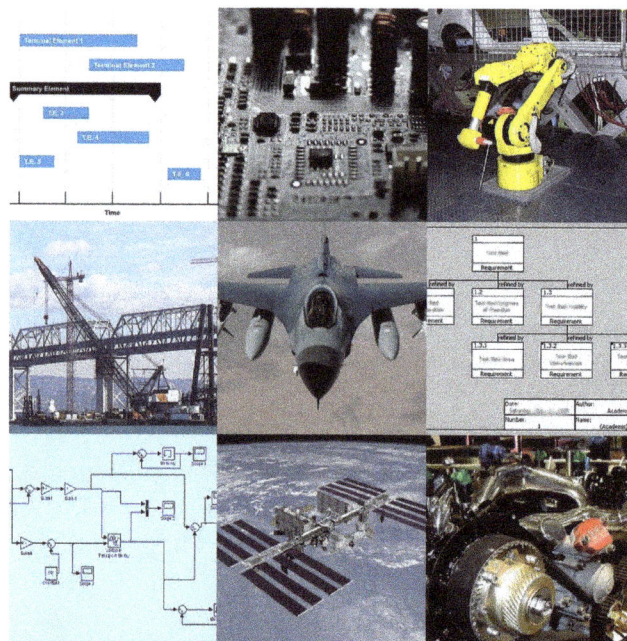

Systems engineering techniques are used in complex projects: spacecraft design, computer chip design, robotics, software integration, and bridge building. Systems engineering uses a host of tools that include modeling and simulation, requirements analysis and scheduling to manage complexity.

Systems engineering is an interdisciplinary field of engineering that focuses on how to design and manage complex engineering systems over their life cycles. Issues such as requirements engineering, reliability, logistics, *coordination* of different teams, testing and evaluation, maintainability and many other disciplines necessary for successful system development, design, implementation, and ultimate decommission become more difficult when dealing with large or complex projects. Systems engineering deals with work-processes, optimization methods, and risk management tools in such projects. It overlaps technical and human-centered disciplines such as industrial engineering, control engineering, software engineering, organizational studies, and project management. Systems engineering ensures that all likely aspects of a project or system are considered, and integrated into a whole.

The systems engineering process is a discovery process that is quite unlike a manufacturing process. A manufacturing process is focused on repetitive activities that achieve high quality outputs with minimum cost and time. The systems engineering process must begin by discovering the real problems that need to be resolved, and identify the most probable or highest impact failures that can occur - systems engineering involves finding elegant solutions to these problems.

History

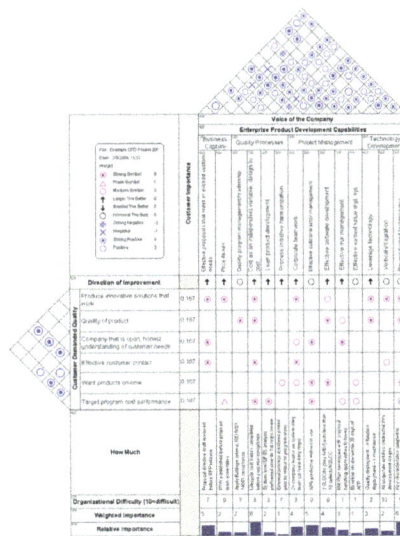

QFD House of Quality for Enterprise Product Development Processes

The term *systems engineering* can be traced back to Bell Telephone Laboratories in the 1940s. The need to identify and manipulate the properties of a system as a whole, which in complex engineering projects may greatly differ from the sum of the parts' properties, motivated various industries, especially those developing systems for the U.S. Military, to apply the discipline.

When it was no longer possible to rely on design evolution to improve upon a system and the existing tools were not sufficient to meet growing demands, new methods began to be developed that addressed the complexity directly. The continuing evolution of systems engineering comprises the development and identification of new methods and modeling techniques. These methods aid in the better comprehension and the design and development control of engineering systems as they grow more complex. Popular tools that are often used in the systems engineering context were developed during these times, including USL, UML, QFD, and IDEF0.

In 1990, a professional society for systems engineering, the *National Council on Systems Engineering* (NCOSE), was founded by representatives from a number of U.S. corporations and organizations. NCOSE was created to address the need for improvements in systems engineering practices and education. As a result of growing involvement from systems engineers outside of the U.S., the name of the organization was changed to the International Council on Systems Engineering

(INCOSE) in 1995. Schools in several countries offer graduate programs in systems engineering, and continuing education options are also available for practicing engineers.

Concept

Some definitions
Simon Ramo considered by some to be a founder of modern systems engineering defined the discipline as:"…a branch of engineering which concentrates on the design and application of the whole as distinct from the parts, looking at a problem in its entirety, taking account of all the facets and all the variables and linking the social to the technological." — Conquering Complexity, 2004.
"An interdisciplinary approach and means to enable the realization of successful systems" — INCOSE handbook, 2004.
"System engineering is a robust approach to the design, creation, and operation of systems. In simple terms, the approach consists of identification and quantification of system goals, creation of alternative system design concepts, performance of design trades, selection and implementation of the best design, verification that the design is properly built and integrated, and post-implementation assessment of how well the system meets (or met) the goals." — NASA Systems Engineering Handbook, 1995.
"The Art and Science of creating effective systems, using whole system, whole life principles" OR "The Art and Science of creating optimal solution systems to complex issues and problems" — Derek Hitchins, Prof. of Systems Engineering, former president of INCOSE (UK), 2007.
"The concept from the engineering standpoint is the evolution of the engineering scientist, i.e., the scientific generalist who maintains a broad outlook. The method is that of the team approach. On large-scale-system problems, teams of scientists and engineers, generalists as well as specialists, exert their joint efforts to find a solution and physically realize it…The technique has been variously called the systems approach or the team development method." — Harry H. Goode & Robert E. Machol, 1957.
"The systems engineering method recognizes each system is an integrated whole even though composed of diverse, specialized structures and sub-functions. It further recognizes that any system has a number of objectives and that the balance between them may differ widely from system to system. The methods seek to optimize the overall system functions according to the weighted objectives and to achieve maximum compatibility of its parts." — Systems Engineering Tools by Harold Chestnut, 1965.

Systems engineering signifies only an approach and, more recently, a discipline in engineering. The aim of education in systems engineering is to formalize various approaches simply and in doing so, identify new methods and research opportunities similar to that which occurs in other fields of engineering. As an approach, systems engineering is holistic and interdisciplinary in flavour.

Origins and Traditional Scope

The traditional scope of engineering embraces the conception, design, development, production and operation of physical systems. Systems engineering, as originally conceived, falls within this scope. "Systems engineering", in this sense of the term, refers to the distinctive set of concepts, methodologies, organizational structures (and so on) that have been developed to meet the challenges of engineering effective functional systems of unprecedented size and complexity within time, budget, and other constraints. The Apollo program is a leading example of a systems engineering project.

Evolution to Broader Scope

The use of the term "systems engineer" has evolved over time to embrace a wider, more holistic concept of "systems" and of engineering processes. This evolution of the definition has been a subject of ongoing controversy, and the term continues to apply to both the narrower and broader scope.

Traditional systems engineering was seen as a branch of engineering in the classical sense, that is, as applied only to physical system, such as space craft and aircraft. More recently, systems engineering has evolved to a take on a broader meaning especially when humans were seen as an essential component of a system. Checkland, for example, captures the broader meaning of systems engineering by stating that 'engineering' "can be read in its general sense; you can engineer a meeting or a political agreement.

Consistent with the broader scope of systems engineering, the Systems Engineering Body of Knowledge (SEBoK) has defined three types of systems engineering: (1) Product Systems Engineering (PSE) is the traditional systems engineering focused on the design of physical systems consisting of hardware and software. (2) Enterprise Systems Engineering (ESE) pertains to the view of enterprises, that is, organizations or combinations of organizations, as systems. (3) Service Systems Engineering (SSE) has to do with the engineering of service systems. Checkland defines a service system as a system which is conceived as serving another system. Most civil infrastructure systems are service systems.

Holistic View

Systems engineering focuses on analyzing and eliciting customer needs and required functionality early in the development cycle, documenting requirements, then proceeding with design synthesis and system validation while considering the complete problem, the system lifecycle. This includes fully understanding all of the stakeholders involved. Oliver et al. claim that the systems engineering process can be decomposed into

- a *Systems Engineering Technical Process*, and

- a *Systems Engineering Management Process*.

Within Oliver's model, the goal of the Management Process is to organize the technical effort in the lifecycle, while the Technical Process includes *assessing available information*, *defining effectiveness measures*, to *create a behavior model*, *create a structure model*, *perform trade-off analysis*, and *create sequential build & test plan*.

Depending on their application, although there are several models that are used in the industry, all of them aim to identify the relation between the various stages mentioned above and incorporate feedback. Examples of such models include the Waterfall model and the VEE model.

Interdisciplinary Field

System development often requires contribution from diverse technical disciplines. By providing a systems (holistic) view of the development effort, systems engineering helps mold all the technical contributors into a unified team effort, forming a structured development process that proceeds from concept to production to operation and, in some cases, to termination and disposal. In an acquisition, the holistic integrative discipline combines contributions and balances tradeoffs among cost, schedule, and performance while maintaining an acceptable level of risk covering the entire life cycle of the item.

This perspective is often replicated in educational programs, in that systems engineering courses are taught by faculty from other engineering departments, which helps create an interdisciplinary environment.

Managing Complexity

The need for systems engineering arose with the increase in complexity of systems and projects, in turn exponentially increasing the possibility of component friction, and therefore the unreliability of the design. When speaking in this context, complexity incorporates not only engineering systems, but also the logical human organization of data. At the same time, a system can become more complex due to an increase in size as well as with an increase in the amount of data, variables, or the number of fields that are involved in the design. The International Space Station is an example of such a system.

The International Space Station is an example of a largely complex system requiring Systems Engineering.

The development of smarter control algorithms, microprocessor design, and analysis of environmental systems also come within the purview of systems engineering. Systems engineering encourages the use of tools and methods to better comprehend and manage complexity in systems. Some examples of these tools can be seen here:

- *System architecture,*

- *System model, Modeling, and Simulation,*

- *Optimization,*

- *System dynamics,*

- *Systems analysis,*

- *Statistical analysis,*

- *Reliability analysis*, and

- *Decision making*

Taking an interdisciplinary approach to engineering systems is inherently complex since the behavior of and interaction among system components is not always immediately well defined or understood. Defining and characterizing such systems and subsystems and the interactions among them is one of the goals of systems engineering. In doing so, the gap that exists between informal requirements from users, operators, marketing organizations, and technical specifications is successfully bridged.

Scope

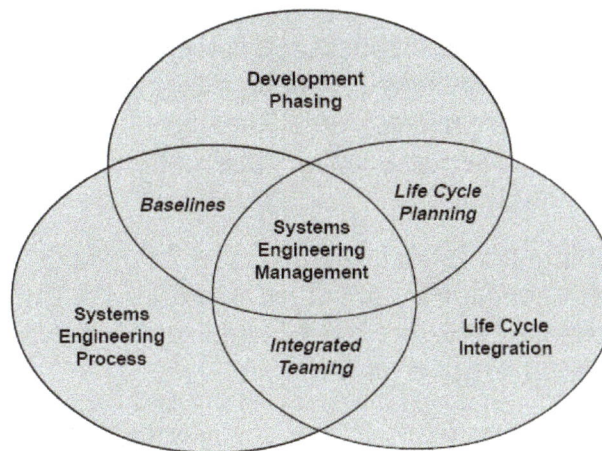

The scope of systems engineering activities

One way to understand the motivation behind systems engineering is to see it as a method, or practice, to identify and improve common rules that exist within a wide variety of systems. Keeping this in mind, the principles of systems engineering – holism, emergent behavior, boundary, et al. – can be applied to any system, complex or otherwise, provided systems thinking is employed at all levels. Besides defense and aerospace, many information and technology based companies, software development firms, and industries in the field of electronics & communications require systems engineers as part of their team.

An analysis by the INCOSE Systems Engineering center of excellence (SECOE) indicates that optimal effort spent on systems engineering is about 15-20% of the total project effort. At the same time, studies have shown that systems engineering essentially leads to reduction in costs among other benefits. However, no quantitative survey at a larger scale encompassing a wide variety of industries has been conducted until recently. Such studies are underway to determine the effectiveness and quantify the benefits of systems engineering.

Systems engineering encourages the use of modeling and simulation to validate assumptions or theories on systems and the interactions within them.

Use of methods that allow early detection of possible failures, in safety engineering, are integrated into the design process. At the same time, decisions made at the beginning of a project whose consequences are not clearly understood can have enormous implications later in the life of a system, and it is the task of the modern systems engineer to explore these issues and make critical decisions. No method guarantees today's decisions will still be valid when a system goes into service years or decades after first conceived. However, there are techniques that support the process of systems engineering. Examples include soft systems methodology, Jay Wright Forrester's System dynamics method, and the Unified Modeling Language (UML)—all currently being explored, evaluated, and developed to support the engineering decision process.

Education

Education in systems engineering is often seen as an extension to the regular engineering courses, reflecting the industry attitude that engineering students need a foundational background in one of the traditional engineering disciplines (e.g., aerospace engineering, civil engineering, electrical engineering, mechanical engineering, industrial engineering)—plus practical, real-world experience to be effective as systems engineers. Undergraduate university programs in systems engineering are rare. Typically, systems engineering is offered at the graduate level in combination with interdisciplinary study.

INCOSE maintains a continuously updated Directory of Systems Engineering Academic Programs worldwide. As of 2009, there are about 80 institutions in United States that offer 165 undergraduate and graduate programs in systems engineering. Education in systems engineering can be taken as *Systems-centric* or *Domain-centric*.

- *Systems-centric* programs treat systems engineering as a separate discipline and most of the courses are taught focusing on systems engineering principles and practice.

- *Domain-centric* programs offer systems engineering as an option that can be exercised with another major field in engineering.

Both of these patterns strive to educate the systems engineer who is able to oversee interdisciplinary projects with the depth required of a core-engineer.

Systems Engineering Topics

Systems engineering tools are strategies, procedures, and techniques that aid in performing systems engineering on a project or product. The purpose of these tools vary from database management, graphical browsing, simulation, and reasoning, to document production, neutral import/ export and more.

System

There are many definitions of what a system is in the field of systems engineering. Below are a few authoritative definitions:

- ANSI/EIA-632-1999: "An aggregation of end products and enabling products to achieve a given purpose."

- DAU Systems Engineering Fundamentals: "an integrated composite of people, products, and processes that provide a capability to satisfy a stated need or objective."

- IEEE Std 1220-1998: "A set or arrangement of elements and processes that are related and whose behavior satisfies customer/operational needs and provides for life cycle sustainment of the products."

- ISO/IEC 15288:2008: "A combination of interacting elements organized to achieve one or more stated purposes."

- NASA Systems Engineering Handbook: "(1) The combination of elements that function together to produce the capability to meet a need. The elements include all hardware, software, equipment, facilities, personnel, processes, and procedures needed for this purpose. (2) The end product (which performs operational functions) and enabling products (which provide life-cycle support services to the operational end products) that make up a system."

- INCOSE Systems Engineering Handbook: "homogeneous entity that exhibits pre-defined behavior in the real world and is composed of heterogeneous parts that do not individually exhibit that behavior and an integrated configuration of components and/ or subsystems."

- INCOSE: "A system is a construct or collection of different elements that together produce results not obtainable by the elements alone. The elements, or parts, can include people, hardware, software, facilities, policies, and documents; that is, all things required to produce systems-level results. The results include system level qualities, properties, characteristics, functions, behavior and performance. The value added by the system as a whole, beyond that contributed independently by the parts, is primarily created by the relationship among the parts; that is, how they are interconnected."

The Systems Engineering Process

Depending on their application, tools are used for various stages of the systems engineering process:

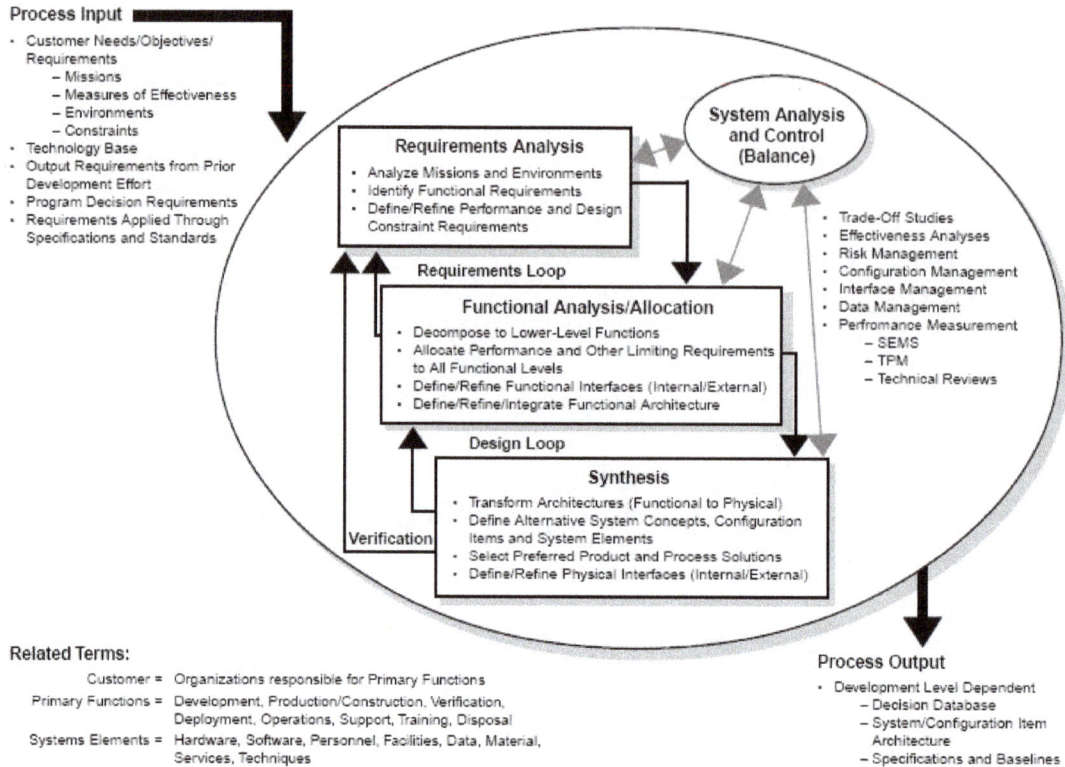

Process Input

- Customer Needs/Objectives/ Requirements
  - Missions
  - Measures of Effectiveness
  - Environments
  - Constraints
- Technology Base
- Output Requirements from Prior Development Effort
- Program Decision Requirements
- Requirements Applied Through Specifications and Standards

Requirements Analysis

- Analyze Missions and Environments
- Identify Functional Requirements
- Define/Refine Performance and Design Constraint Requirements

System Analysis and Control (Balance)

- Trade-Off Studies
- Effectiveness Analyses
- Risk Management
- Configuration Management
- Interface Management
- Data Management
- Perfromance Measurement
  - SEMS
  - TPM
  - Technical Reviews

Requirements Loop

Functional Analysis/Allocation

- Decompose to Lower-Level Functions
- Allocate Performance and Other Limiting Requirements to All Functional Levels
- Define/Refine Functional Interfaces (Internal/External)
- Define/Refine/Integrate Functional Architecture

Design Loop

Synthesis

- Transform Architectures (Functional to Physical)
- Define Alternative System Concepts, Configuration Items and System Elements
- Select Preferred Product and Process Solutions
- Define/Refine Physical Interfaces (Internal/External)

Verification

Related Terms:

Customer = Organizations responsible for Primary Functions
Primary Functions = Development, Production/Construction, Verification, Deployment, Operations, Support, Training, Disposal
Systems Elements = Hardware, Software, Personnel, Facilities, Data, Material, Services, Techniques

Process Output

- Development Level Dependent
  - Decision Database
  - System/Configuration Item Architecture
  - Specifications and Baselines

Using Models

Models play important and diverse roles in systems engineering. A model can be defined in several ways, including:

- An abstraction of reality designed to answer specific questions about the real world

- An imitation, analogue, or representation of a real world process or structure; or

- A conceptual, mathematical, or physical tool to assist a decision maker.

Together, these definitions are broad enough to encompass physical engineering models used in the verification of a system design, as well as schematic models like a functional flow block diagram and mathematical (i.e., quantitative) models used in the trade study process.

The main reason for using mathematical models and diagrams in trade studies is to provide estimates of system effectiveness, performance or technical attributes, and cost from a set of known or estimable quantities. Typically, a collection of separate models is needed to provide all of these outcome variables. The heart of any mathematical model is a set of meaningful quantitative relationships among its inputs and outputs. These relationships can be as simple as adding up constituent quantities to obtain a total, or as complex as a set of differential equations describing the trajectory of a spacecraft in a gravitational field. Ideally, the relationships express causality, not just correlation. Furthermore, key to successful systems engineering activities are also the methods with which these models are efficiently and effectively managed and used to simulate the systems. However, diverse domains often present recurring problems of modeling and simulation for systems engineering, and new advancements are aiming to crossfertilize methods among distinct scientific and engineering communities, under the title of 'Modeling & Simulation-based Systems Engineering'.

Modeling Formalisms and Graphical Representations

Initially, when the primary purpose of a systems engineer is to comprehend a complex problem, graphic representations of a system are used to communicate a system's functional and data requirements. Common graphical representations include:

- Functional flow block diagram (FFBD)

- Model-based design, for example Simulink, VisSim, VisualSim etc.

- Data Flow Diagram (DFD)

- N2 Chart

- IDEF0 Diagram

- Use case diagram

- Sequence diagram

- Block diagram

- Signal-flow graph

- USL Function Maps and Type Maps.

- Enterprise Architecture frameworks, like TOGAF, MODAF, Zachman Frameworks etc.

A graphical representation relates the various subsystems or parts of a system through functions, data, or interfaces. Any or each of the above methods are used in an industry based on its requirements. For instance, the N2 chart may be used where interfaces between systems is important. Part of the design phase is to create structural and behavioral models of the system.

Once the requirements are understood, it is now the responsibility of a systems engineer to refine them, and to determine, along with other engineers, the best technology for a job. At this point starting with a trade study, systems engineering encourages the use of weighted choices to determine the best option. A decision matrix, or Pugh method, is one way (QFD is another) to make this choice while considering all criteria that are important. The trade study in turn informs the design, which again affects graphic representations of the system (without changing the requirements). In an SE process, this stage represents the iterative step that is carried out until a feasible solution is found. A decision matrix is often populated using techniques such as statistical analysis, reliability analysis, system dynamics (feedback control), and optimization methods.

Other Tools

Systems Modeling Language (SysML), a modeling language used for systems engineering applications, supports the specification, analysis, design, verification and validation of a broad range of complex systems.

Lifecycle Modeling Language (LML), is an open-standard modeling language designed for systems engineering that supports the full lifecycle: conceptual, utilization, support and retirement stages.

Related Fields and Sub-fields

Many related fields may be considered tightly coupled to systems engineering. These areas have contributed to the development of systems engineering as a distinct entity.

Cognitive Systems Engineering

> Cognitive systems engineering (CSE) is a specific approach to the description and analysis of human-machine systems or sociotechnical systems. The three main themes of CSE are how humans cope with complexity, how work is accomplished by the use of artifacts, and how human-machine systems and socio-technical systems can be described as joint cognitive systems. CSE has since its beginning become a recognized scientific discipline, sometimes also referred to as cognitive engineering. The concept of a Joint Cognitive System (JCS) has in particular become widely used as a way of understanding how complex socio-technical systems can be described with varying degrees of resolution. The more than 20 years of experience with CSE has been described extensively.

Configuration management

> Like systems engineering, configuration management as practiced in the defense and aerospace industry is a broad systems-level practice. The field parallels the taskings of systems engineering; where systems engineering deals with requirements development, allocation to development items and verification, configuration management deals with requirements capture, traceability to the development item, and audit of development item to ensure that it has achieved the desired functionality that systems engineering and/ or Test and Verification Engineering have proven out through objective testing.

Control engineering

> Control engineering and its design and implementation of control systems, used extensively in nearly every industry, is a large sub-field of systems engineering. The cruise control on an automobile and the guidance system for a ballistic missile are two examples. Control systems theory is an active field of applied mathematics involving the investigation of solution spaces and the development of new methods for the analysis of the control process.

Industrial engineering

Industrial engineering is a branch of engineering that concerns the development, improvement, implementation and evaluation of integrated systems of people, money, knowledge, information, equipment, energy, material and process. Industrial engineering draws upon the principles and methods of engineering analysis and synthesis, as well as mathematical, physical and social sciences together with the principles and methods of engineering analysis and design to specify, predict, and evaluate results obtained from such systems.

Interface design

Interface design and its specification are concerned with assuring that the pieces of a system connect and inter-operate with other parts of the system and with external systems as necessary. Interface design also includes assuring that system interfaces be able to accept new features, including mechanical, electrical and logical interfaces, including reserved wires, plug-space, command codes and bits in communication protocols. This is known as extensibility. Human-Computer Interaction (HCI) or Human-Machine Interface (HMI) is another aspect of interface design, and is a critical aspect of modern systems engineering. Systems engineering principles are applied in the design of network protocols for local-area networks and wide-area networks.

Mechatronic engineering

Mechatronic engineering, like systems engineering, is a multidisciplinary field of engineering that uses dynamical systems modeling to express tangible constructs. In that regard it is almost indistinguishable from Systems Engineering, but what sets it apart is the focus on smaller details rather than larger generalizations and relationships. As such, both fields are distinguished by the scope of their projects rather than the methodology of their practice.

Operations research

Operations research supports systems engineering. The tools of operations research are used in systems analysis, decision making, and trade studies. Several schools teach SE courses within the operations research or industrial engineering department, highlighting the role systems engineering plays in complex projects. Operations research, briefly, is concerned with the optimization of a process under multiple constraints.

Performance engineering

Performance engineering is the discipline of ensuring a system meets customer expectations for performance throughout its life. Performance is usually defined as the speed with which a certain operation is executed, or the capability of executing a number of such operations in a unit of time. Performance may be degraded when an operations queue to execute is throttled by limited system capacity. For example, the performance of a packet-switched network is characterized by the end-to-end packet transit delay, or the number of packets switched in an hour. The design of high-performance systems uses analytical or simulation modeling, whereas the delivery of high-performance implementation involves thorough performance testing. Performance engineering relies heavily on statistics, queueing theory and probability theory for its tools and processes.

Program management and project management.

Program management (or programme management) has many similarities with systems engineering, but has broader-based origins than the engineering ones of systems engineering. Project management is also closely related to both program management and systems engineering.

Proposal engineering

Proposal engineering is the application of scientific and mathematical principles to design, construct, and operate a cost-effective proposal development system. Basically, proposal engineering uses the "systems engineering process" to create a cost effective proposal and increase the odds of a successful proposal.

Reliability engineering

Reliability engineering is the discipline of ensuring a system meets customer expectations for reliability throughout its life; i.e., it does not fail more frequently than expected. Reliability engineering applies to all aspects of the system. It is closely associated with maintainability, availability (dependability or RAMS preferred by some), and logistics engineering. Reliability engineering is always a critical component of safety engineering, as in failure modes and effects analysis (FMEA) and hazard fault tree analysis, and of security engineering.

Risk Management

Risk Management, the practice of assessing and dealing with risk is one of the interdisciplinary parts of Systems Engineering. In development, acquisition, or operational activities, the inclusion of risk in tradeoff with cost, schedule, and performance features, involves the iterative complex configuration management of traceability and evaluation to the scheduling and requirements management across domains and for the system lifecycle that requires the interdisciplinary technical approach of systems engineering.

Safety engineering

The techniques of safety engineering may be applied by non-specialist engineers in designing complex systems to minimize the probability of safety-critical failures. The "System Safety Engineering" function helps to identify "safety hazards" in emerging designs, and may assist with techniques to "mitigate" the effects of (potentially) hazardous conditions that cannot be designed out of systems.

Scheduling

Scheduling is one of the systems engineering support tools as a practice and item in assessing interdisciplinary concerns under configuration management. In particular the direct relationship of resources, performance features, and risk to duration of a task or the dependency links among tasks and impacts across the system lifecycle are systems engineering concerns.

Security engineering

Security engineering can be viewed as an interdisciplinary field that integrates the community of practice for control systems design, reliability, safety and systems engineering. It may involve such sub-specialties as authentication of system users, system targets and others: people, objects and processes.

Software engineering

From its beginnings, software engineering has helped shape modern systems engineering practice. The techniques used in the handling of complexes of large software-intensive systems have had a major effect on the shaping and reshaping of the tools, methods and processes of SE.

Manufacturing Engineering

Manufacturing engineering is a discipline of engineering dealing with different manufacturing practices and includes the research, design and development of systems, processes, machines, tools, and equipment. The manufacturing engineer's primary focus is to turn raw materials into a new or updated product in the most economic, efficient, and effective way possible.

Overview

This field also deals with the integration of different facilities and systems for producing quality products (with optimal expenditure) by applying the principles of physics and the results of manufacturing systems studies, such as the following:

• Craft or Guild	• Computer integrated manufacturing	• Agile manufacturing
• Putting-out system	• Computer-aided technologies in manufacturing	• Rapid manufacturing
• English system of manufacturing	• Just in time manufacturing	• Prefabrication
• American system of manufacturing	• Lean manufacturing	• Ownership
• Soviet collectivism in manufacturing	• Flexible manufacturing	• Fabrication
• Mass production	• Mass customization	• Publication

Manufacturing engineers develop and create physical artifacts, production processes, and technology. It is a very broad area which includes the design and development of products. Manufacturing engineering is considered to be a subdiscipline of industrial engineering/systems engineering and has very strong overlaps with mechanical engineering. Manufacturing engineers' success or failure directly impacts the advancement of technology and the spread of innovation.

This field of manufacturing engineering emerged from tool and die discipline in the early 20th century. It expanded greatly from the 1960s when industrialized countries introduced factories with:

1. Numerical control machine tools and automated systems of production.

2. Advanced statistical methods of quality control: These factories were pioneered by the American electrical engineer William Edwards Deming, who was initially ignored by his home country. The same methods of quality control later turned Japanese factories into world leaders in cost-effectiveness and production quality.

3. Industrial robots on the factory floor, introduced in the late 1970s: These computer-controlled welding arms and grippers could perform simple tasks such as attaching a car door quickly and flawlessly 24 hours a day. This cut costs and improved production speed.

History

The history of manufacturing engineering can be traced to factories in the mid 19th century USA and 18th century UK. Although large home production sites and workshops were established in ancient China, ancient Rome and the Middle East, the Venice Arsenal provides one of the first examples of a factory in the modern sense of the word. Founded in 1104 in the Republic of Venice several hundred years before the Industrial Revolution, this factory mass-produced ships on assembly lines using manufactured parts. The Venice Arsenal apparently produced nearly one ship every day and, at its height, employed 16,000 people.

Many historians regard Matthew Boulton's Soho Manufactory (established in 1761 in Birmingham) as the first modern factory. Similar claims can be made for John Lombe's silk mill in Derby (1721), or Richard Arkwright's Cromford Mill (1771). The Cromford Mill was purpose-built to accommodate the equipment it held and to take the material through the various manufacturing processes.

One historian, Murno Gladst, contends that the first factory was in Potosí. The Potosi factory took advantage of the abundant silver that was mined nearby and processed silver ingot slugs into coins.

British colonies in the 19th century built factories simply as buildings where a large number of workers gathered to perform hand labor, usually in textile production. This proved more efficient for the administration and distribution of materials to individual workers than earlier methods of manufacturing, such as cottage industries or the putting-out system.

Cotton mills used inventions such as the steam engine and the power loom to pioneer the industrial factories of the 19th century, where precision machine tools and replaceable parts allowed greater efficiency and less waste. This experience formed the basis for the later studies of manufacturing engineering. Between 1820 and 1850, non-mechanized factories supplanted traditional artisan shops as the predominant form of manufacturing institution.

Henry Ford further revolutionized the factory concept and thus manufacturing engineering in the early 20th century with the innovation of mass production. Highly specialized workers situated alongside a series of rolling ramps would build up a product such as (in Ford's case) an automobile. This concept dramatically decreased production costs for virtually all manufactured goods and brought about the age of consumerism.

Modern Developments

Modern manufacturing engineering studies include all intermediate processes required for the production and integration of a product's components.

Some industries, such as semiconductor and steel manufacturers use the term "fabrication" for these processes.

Automation is used in different processes of manufacturing such as machining and welding. Automated manufacturing refers to the application of automation to produce goods in a factory. The main advantages of automated manufacturing for the manufacturing process are realized with effective implementation of automation and include: higher consistency and quality, reduction of lead times, simplification of production, reduced handling, improved work flow, and improved worker morale.

Robotics is the application of mechatronics and automation to create robots, which are often used in manufacturing to perform tasks that are dangerous, unpleasant, or repetitive. These robots may be of any shape and size, but all are preprogrammed and interact physically with the world. To create a robot, an engineer typically employs kinematics (to determine the robot's range of motion) and mechanics (to determine the stresses within the robot). Robots are used extensively in manufacturing engineering.

Robots allow businesses to save money on labor, perform tasks that are either too dangerous or too precise for humans to perform economically, and to ensure better quality. Many companies employ assembly lines of robots, and some factories are so robotized that they can run by themselves. Outside the factory, robots have been employed in bomb disposal, space exploration, and many other fields. Robots are also sold for various residential applications.

Education

Certification Programs

Manufacturing engineers possess a bachelor's degree in engineering with a major in manufacturing engineering. The length of study for such a degree is usually four to five years followed by five more years of professional practice to qualify as a professional engineer. Working as a manufacturing engineering technologist involves a more applications-oriented qualification path.

Academic degrees for manufacturing engineers are usually the Bachelor of Engineering, [BE] or [BEng], and the Bachelor of Science, [BS] or [BSc]. For manufacturing technologists the required degrees are Bachelor of Technology [B.TECH] or Bachelor of Applied Science [BASc] in Manufacturing, depending upon the university. Master's degrees in engineering manufacturing include Master of Engineering [ME] or [MEng] in Manufacturing, Master of Science [M.Sc] in Manufacturing Management, Master of Science [M.Sc] in Industrial and Production Management, and Master of Science [M.Sc] as well as Master of Engineering [ME] in Design, which is a subdiscipline of manufacturing.

Doctoral [PhD] or [DEng] level courses in manufacturing are also available depending on the university.

The undergraduate degree curriculum generally includes courses in physics, mathematics, computer science, project management, and specific topics in mechanical and manufacturing engineering. Initially such topics cover most, if not all, of the subdisciplines of manufacturing engineering. Students then choose to specialize in one or more subdisciplines towards the end of their degree work.

Syllabus

The foundational curriculum for a bachelor's degree in manufacturing engineering is very similar to that for mechanical engineering, and includes:

- Statics and dynamics

- Strength of materials and solid mechanics

- Instrumentation and measurement

- Applied thermodynamics, heat transfer, energy conversion, and HVAC

- Fluid mechanics and fluid dynamics

- Mechanism design (including kinematics and dynamics)

- Manufacturing technology or processes

- Hydraulics and pneumatics

- Mathematics - in particular, calculus, differential equations, statistics, and linear algebra.

- Engineering design and graphics

- Circuit Analysis

- Lean manufacturing

- Mechatronics and control theory

- Automation and reverse engineering

- Quality assurance and control

- Material science

- Drafting, CAD (including solid modeling), and CAM, etc.

A bachelor's degree in these two areas will typically differ only by a few specialized classes, although the mechanical engineering degree requires more mathematics expertise.

Manufacturing Engineering Certification

Certification and licensure:

In some countries, "professional engineer" is the term for registered or licensed engineers who are permitted to offer their professional services directly to the public. Professional Engineer, abbreviated (PE - USA) or (PEng - Canada), is the designation for licensure in North America. In order to qualify for this license, a candidate needs a bachelor's degree from an ABET recognized university in the USA, a passing score on a state examination, and four years of work experience usually gained via a structured internship. In the USA, more recent graduates have the option of dividing this licensure process into two segments. The Fundamentals of Engineering (FE) exam is often taken immediately after graduation and the Principles and Practice of Engineering exam is taken after four years of working in a chosen engineering field.

Society of Manufacturing Engineers (SME) certifications (USA):

The SME administers qualifications specifically for the manufacturing industry. These are not degree level qualifications and are not recognized at the professional engineering level. The following discussion deals with qualifications in the USA only. Qualified candidates for the Certified Manufacturing Technologist Certificate (CMfgT) must pass a three-hour, 130-question multiple-choice exam. The exam covers math, manufacturing processes, manufacturing management, automation, and related subjects. Additionally, a candidate must have at least four years of combined education and manufacturing-related work experience.

Certified Manufacturing Engineer (CMfgE) is an engineering qualification administered by the Society of Manufacturing Engineers, Dearborn, Michigan, USA. Candidates qualifying for a Certified Manufacturing Engineer credential must pass a four-hour, 180 question multiple-choice exam which covers more in-depth topics than does the CMfgT exam. CMfgE candidates must also have eight years of combined education and manufacturing-related work experience, with a minimum of four years of work experience.

Certified Engineering Manager (CEM). The Certified Engineering Manager Certificate is also designed for engineers with eight years of combined education and manufacturing experience. The test is four hours long and has 160 multiple-choice questions. The CEM certification exam covers business processes, teamwork, responsibility, and other management-related categories.

Many manufacturing companies, especially those in industrialized nations, have begun to incorporate computer-aided engineering (CAE) programs into their existing design and analysis processes, including 2D and 3D solid modeling computer-aided design (CAD). This method has many benefits, including easier and more exhaustive visualization of products, the ability to create virtual assemblies of parts, and ease of use in designing mating interfaces and tolerances.

Other CAE programs commonly used by product manufacturers include product life cycle management (PLM) tools and analysis tools used to perform complex simulations. Analysis tools may be used to predict product response to expected loads, including fatigue life and manufacturability. These tools include finite element analysis (FEA), computational fluid dynamics (CFD), and computer-aided manufacturing (CAM).

Using CAE programs, a mechanical design team can quickly and cheaply iterate the design process to develop a product that better meets cost, performance, and other constraints. No physical prototype need be created until the design nears completion, allowing hundreds or thousands of designs to be evaluated, instead of relatively few. In addition, CAE analysis programs can model complicated physical phenomena which cannot be solved by hand, such as viscoelasticity, complex contact between mating parts, or non-Newtonian flows.

Just as manufacturing engineering is linked with other disciplines, such as mechatronics, multidisciplinary design optimization (MDO) is also being used with other CAE programs to automate and improve the iterative design process. MDO tools wrap around existing CAE processes, allowing product evaluation to continue even after the analyst goes home for the day. They also utilize sophisticated optimization algorithms to more intelligently explore possible designs, often finding better, innovative solutions to difficult multidisciplinary design problems.

Mechanics, in the most general sense, is the study of forces and their effects on matter. Typically, engineering mechanics is used to analyze and predict the acceleration and deformation (both elastic and plastic) of objects under known forces (also called loads) or stresses. Subdisciplines of mechanics include:

- Statics, the study of non-moving bodies under known loads

- Dynamics (or kinetics), the study of how forces affect moving bodies

- Mechanics of materials, the study of how different materials deform under various types of stress

- Fluid mechanics, the study of how fluids react to forces

- Continuum mechanics, a method of applying mechanics that assumes that objects are continuous (rather than discrete)

If the engineering project were to design a vehicle, statics might be employed to design the frame of the vehicle in order to evaluate where the stresses will be most intense. Dynamics might be used when designing the car's engine to evaluate the forces in the pistons and cams as the engine cycles. Mechanics of materials might be used to choose appropriate materials for the manufacture of the frame and engine. Fluid mechanics might be used to design a ventilation system for the vehicle or to design the intake system for the engine.

Kinematics

Kinematics is the study of the motion of bodies (objects) and systems (groups of objects), while ignoring the forces that cause the motion. The movement of a crane and the oscillations of a piston in an engine are both simple kinematic systems. The crane is a type of open kinematic chain, while the piston is part of a closed four-bar linkage. Engineers typically use kinematics in the design and analysis of mechanisms. Kinematics can be used to find the possible range of motion for a given mechanism, or, working in reverse, can be used to design a mechanism that has a desired range of motion.

Drafting or technical drawing is the means by which manufacturers create instructions for manufacturing parts. A technical drawing can be a computer model or hand-drawn schematic showing all the dimensions necessary to manufacture a part, as well as assembly notes, a list of required materials, and other pertinent information. A U.S engineer or skilled worker who creates technical drawings may be referred to as a drafter or draftsman. Drafting has historically been a two-dimensional process, but computer-aided design (CAD) programs now allow the designer to create in three dimensions.

Instructions for manufacturing a part must be fed to the necessary machinery, either manually, through programmed instructions, or through the use of a computer-aided manufacturing (CAM) or combined CAD/CAM program. Optionally, an engineer may also manually manufacture a part using the technical drawings, but this is becoming an increasing rarity with the advent of computer numerically controlled (CNC) manufacturing. Engineers primarily manufacture parts manually in the areas of applied spray coatings, finishes, and other processes that cannot economically or practically be done by a machine.

Drafting is used in nearly every subdiscipline of mechanical and manufacturing engineering, and by many other branches of engineering and architecture. Three-dimensional models created using CAD software are also commonly used in finite element analysis (FEA) and computational fluid dynamics (CFD).

Mechatronics is an engineering discipline that deals with the convergence of electrical, mechanical and manufacturing systems. Such combined systems are known as electromechanical systems and are widespread. Examples include automated manufacturing systems, heating, ventilation and air-conditioning systems, and various aircraft and automobile subsystems.

The term mechatronics is typically used to refer to macroscopic systems, but futurists have predicted the emergence of very small electromechanical devices. Already such small devices, known as Microelectromechanical systems (MEMS), are used in automobiles to initiate the deployment of airbags, in digital projectors to create sharper images, and in inkjet printers to create nozzles for high-definition printing. In the future it is hoped that such devices will be used in tiny implantable medical devices and to improve optical communication.

Textile Engineering

Textile engineering courses deal with the application of scientific and engineering principles to the design and control of all aspects of fiber, textile, and apparel processes, products, and machinery. These include natural and man-made materials, interaction of materials with machines, safety and health, energy conservation, and waste and pollution control. Additionally, students are given experience in plant design and layout, machine and wet process design and improvement, and designing and creating textile products. Throughout the textile engineering curriculum, students take classes from other engineering and disciplines including: mechanical, chemical, materials and industrial engineering.

Employment

Manufacturing engineering is just one facet of the engineering industry. Manufacturing engineers enjoy improving the production process from start to finish. They have the ability to keep the whole production process in mind as they focus on a particular portion of the process. Successful students in manufacturing engineering degree programs are inspired by the notion of starting with a natural resource, such as a block of wood, and ending with a usable, valuable product, such as a desk, produced efficiently and economically.

Manufacturing engineers are closely connected with engineering and industrial design efforts. Examples of major companies that employ manufacturing engineers in the United States include General Motors Corporation, Ford Motor Company, Chrysler, Boeing, Gates Corporation and Pfizer. Examples in Europe include Airbus, Daimler, BMW, Fiat, Navistar International, and Michelin Tyre.

Industries where manufacturing engineers are generally employed include:

- Aerospace industry
- Automotive industry
- Chemical industry
- Computer industry
- Food processing industry
- Garment industry
- Pharmaceutical industry
- Pulp and paper industry
- Toy industry

A flexible manufacturing system (FMS) is a manufacturing system in which there is some amount of flexibility that allows the system to react to changes, whether predicted or unpredicted. This flexibility is generally considered to fall into two categories, both of which have numerous subcategories. The first category, machine flexibility, covers the system's ability to be changed to produce new product types and the ability to change the order of operations executed on a part. The second category, called routing flexibility, consists of the ability to use multiple machines to perform the same operation on a part, as well as the system's ability to absorb large-scale changes, such as in volume, capacity, or capability.

Most FMS systems comprise three main systems. The work machines, which are often automated CNC machines, are connected by a material handling system to optimize parts flow, and to a central control computer, which controls material movements and machine flow. The main advantages of an FMS is its high flexibility in managing manufacturing resources like time and effort in order to manufacture a new product. The best application of an FMS is found in the production of small sets of products from a mass production.

Computer Integrated Manufacturing

Computer-integrated manufacturing (CIM) in engineering is a method of manufacturing in which the entire production process is controlled by computer. Traditionally separated process methods are joined through a computer by CIM. This integration allows the processes to exchange information and to initiate actions. Through this integration, manufacturing can be faster and less error-prone, although the main advantage is the ability to create automated manufacturing processes. Typically CIM relies on closed-loop control processes based on real-time input from sensors. It is also known as flexible design and manufacturing.

Friction stir welding was discovered in 1991 by The Welding Institute (TWI). This innovative steady state (non-fusion) welding technique joins previously un-weldable materials, including several aluminum alloys. It may play an important role in the future construction of airplanes, potentially replacing rivets. Current uses of this technology to date include: welding the seams of the aluminum main space shuttle external tank, the Orion Crew Vehicle test article, Boeing Delta II and Delta IV Expendable Launch Vehicles and the SpaceX Falcon 1 rocket; armor plating for amphibious assault ships; and welding the wings and fuselage panels of the new Eclipse 500 aircraft from Eclipse Aviation, among an increasingly growing range of uses.

Other areas of research are Product Design, MEMS (Micro-Electro-Mechanical Systems), Lean Manufacturing, Intelligent Manufacturing Systems, Green Manufacturing, Precision Engineering, Smart Materials, etc.

Human Factors and Ergonomics

Human factors and ergonomics (HF&E), also known as comfort design, functional design, and systems, is the practice of designing products, systems, or processes to take proper account of the interaction between them and the people who use them.

The field has seen contributions from numerous disciplines, such as psychology, engineering, biomechanics, industrial design, physiology, and anthropometry. In essence, it is the study of designing equipment, devices and processes that fit the human body and its cognitive abilities. The two terms "human factors" and "ergonomics" are essentially synonymous.

The International Ergonomics Association defines ergonomics or human factors as follows:

Ergonomics (or human factors) is the scientific discipline concerned with the understanding of interactions among humans and other elements of a system, and the profession that applies theory, principles, data and methods to design in order to optimize human well-being and overall system performance.

HF&E is employed to fulfill the goals of occupational health and safety and productivity. It is relevant in the design of such things as safe furniture and easy-to-use interfaces to machines and equipment.

Proper ergonomic design is necessary to prevent repetitive strain injuries and other musculoskeletal disorders, which can develop over time and can lead to long-term disability.

Human factors and ergonomics is concerned with the "fit" between the user, equipment and their environments. It takes account of the user's capabilities and limitations in seeking to ensure that tasks, functions, information and the environment suit each user.

To assess the fit between a person and the used technology, human factors specialists or ergonomists consider the job (activity) being done and the demands on the user; the equipment used (its size, shape, and how appropriate it is for the task), and the information used (how it is presented, accessed, and changed). Ergonomics draws on many disciplines in its study of humans and their environments, including anthropometry, biomechanics, mechanical engineering, industrial engineering, industrial design, information design, kinesiology, physiology, cognitive psychology, industrial and organizational psychology, and space psychology.

Etymology

The term *ergonomics* first entered the modern lexicon when Polish scientist Wojciech Jastrzębowski used the word in his 1857 article *Rys ergonomji czyli nauki o pracy, opartej na prawdach poczerpniętych z Nauki Przyrody* (The Outline of Ergonomics; i.e. Science of Work, Based on the Truths Taken from the Natural Science). The introduction of the term to the English lexicon is widely attributed to British psychologist Hywel Murrell, at the 1949 meeting at the UK's Admiralty, which led to the foundation of The Ergonomics Society. He used it to encompass the studies in which he had been engaged during and after World War II.

The expression *human factors* is a predominantly North American term which has been adopted to emphasise the application of the same methods to non work-related situations. A "human factor" is a physical or cognitive property of an individual or social behavior specific to humans that may influence the functioning of technological systems. The terms "human factors" and "ergonomics" are essentially synonymous.

Domains of Specialization

Ergonomics comprise three main fields of research: Physical, cognitive and organisational ergonomics.

There are many specializations within these broad categories. Specialisations in the field of physical ergonomics may include visual ergonomics. Specialisations within the field of cognitive ergonomics may include usability, human–computer interaction, and user experience engineering.

Some specialisations may cut across these domains: *Environmental ergonomics* is concerned with human interaction with the environment as characterized by climate, temperature, pressure, vibration, light. The emerging field of human factors in highway safety uses human factor principles to understand the actions and capabilities of road users - car and truck drivers, pedestrians, bicyclists, etc. - and use this knowledge to design roads and streets to reduce traffic collisions. Driver error is listed as a contributing factor in 44% of fatal collisions in the United States, so a topic of particular interest is how road users gather and process information about the road and its environment, and how to assist them to make the appropriate decision.

New terms are being generated all the time. For instance, "user trial engineer" may refer to a human factors professional who specialises in user trials. Although the names change, human factors professionals apply an understanding of human factors to the design of equipment, systems and working methods in order to improve comfort, health, safety, and productivity.

According to the International Ergonomics Association, within the discipline of ergonomics there exist domains of specialization:

Physical Ergonomics

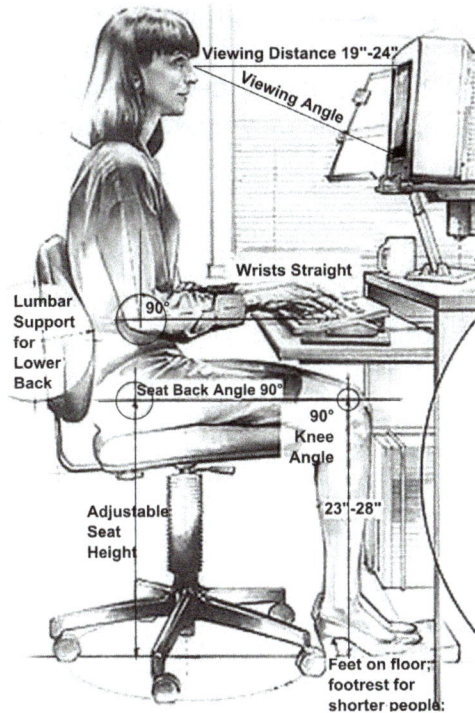

Physical ergonomics: the science of designing user interaction with equipment and workplaces to fit the user.

Physical ergonomics is concerned with human anatomy, and some of the anthropometric, physiological and bio mechanical characteristics as they relate to physical activity. Physical ergonomic principles have been widely used in the design of both consumer and industrial products. Physical ergonomics is important in the medical field, particularly to those diagnosed with physiological ailments or disorders such as arthritis (both chronic and temporary) or carpal tunnel syndrome. Pressure that is insignificant or imperceptible to those unaffected by these disorders may be very painful, or render a device unusable, for those who are. Many ergonomically designed products are also used or recommended to treat or prevent such disorders, and to treat pressure-related chronic pain.

One of the most prevalent types of work-related injuries is musculoskeletal disorder. Work-related musculoskeletal disorders (WRMDs) result in persistent pain, loss of functional capacity and work disability, but their initial diagnosis is difficult because they are mainly based on complaints of pain and other symptoms. Every year, 1.8 million U.S. workers experience WRMDs and nearly 600,000 of the injuries are serious enough to cause workers to miss work. Certain jobs or work conditions cause a higher rate of worker complaints of undue strain, localized fatigue, discomfort, or pain that does not go away after overnight rest. These types of jobs are often those involving activities such as repetitive and forceful exertions; frequent, heavy, or overhead lifts; awkward work positions; or use of vibrating equipment. The Occupational Safety and Health Administration (OSHA) has found substantial evidence that ergonomics programs can cut workers' compensation costs, increase productivity and decrease employee turnover. Therefore, it is important to gather data to identify jobs or work conditions that are most problematic, using sources such as injury and illness logs, medical records, and job analyses.

Cognitive Ergonomics

Cognitive ergonomics is concerned with mental processes, such as perception, memory, reasoning, and motor response, as they affect interactions among humans and other elements of a system. (Relevant topics include mental workload, decision-making, skilled performance, human reliability, work stress and training as these may relate to human-system and Human-Computer Interaction design.)

Organizational Ergonomics

Organizational ergonomics is concerned with the optimization of socio-technical systems, including their organizational structures, policies, and processes. (Relevant topics include communication, crew resource management, work design, work systems, design of working times, teamwork, participatory design, community ergonomics, cooperative work, new work programs, virtual organizations, telework, and quality management.)

History of the Field

In Ancient Societies

The foundations of the science of ergonomics appear to have been laid within the context of the culture of Ancient Greece. A good deal of evidence indicates that Greek civilization in the 5th century BC used ergonomic principles in the design of their tools, jobs, and workplaces. One outstanding example of this can be found in the description Hippocrates gave of how a surgeon's workplace should be designed and how the tools he uses should be arranged. The archaeological record also shows that the early Egyptian dynasties made tools and household equipment that illustrated ergonomic principles.

In Industrial Societies

In the 19th century, Frederick Winslow Taylor pioneered the "scientific management" method, which proposed a way to find the optimum method of carrying out a given task. Taylor found that he could, for example, triple the amount of coal that workers were shoveling by incrementally reducing the size and weight of coal shovels until the fastest shoveling rate was reached. Frank and Lillian Gilbreth expanded Taylor's methods in the early 1900s to develop the "time and motion study". They aimed to improve efficiency by eliminating unnecessary steps and actions. By applying this approach, the Gilbreths reduced the number of motions in bricklaying from 18 to 4.5, allowing bricklayers to increase their productivity from 120 to 350 bricks per hour.

However, this approach was rejected by Russian researchers who focused on the well being of the worker. At the First Conference on Scientific Organization of Labour (1921) Vladimir Bekhterev and Vladimir Nikolayevich Myasishchev criticised Taylorism. Bekhterev argued that "The ultimate ideal of the labour problem is not in it [Taylorism], but is in such organisation of the labour process that would yield a maximum of efficiency coupled with a minimum of health hazards, absence of fatigue and a guarantee of the sound health and all round personal development of the working people." Myasishchev rejected Frederick Taylor's proposal to turn man into a machine. Dull monotonous work was a temporary necessity until a corresponding machine can be developed. He also went on to suggest a new discipline of "ergology" to study work as an integral part of the re-organisation of work. The concept was taken up by Myasishchev's mentor, Bekhterev, in his final report on the conference, merely changing the name to "ergonology"

In Aviation

Prior to World War I, the focus of aviation psychology was on the aviator himself, but the war shifted the focus onto the aircraft, in particular, the design of controls and displays, and the effects of altitude and environmental factors on the pilot. The war saw the emergence of aeromedical research and the need for testing and measurement methods. Studies on driver behaviour started gaining momentum during this period, as Henry Ford started providing millions of Americans with automobiles. Another major development during this period was the performance of aeromedical research. By the end of World War I, two aeronautical labs were established, one at Brooks Air Force Base, Texas and the other at Wright-Patterson Air Force Base outside of Dayton, Ohio. Many tests were conducted to determine which characteristic differentiated the successful pilots from the unsuccessful ones. During the early 1930s, Edwin Link developed the first flight simulator. The trend continued and more sophisticated simulators and test equipment were developed. Another significant development was in the civilian sector, where the effects of illumination on worker productivity were examined. This led to the identification of the Hawthorne Effect, which suggested that motivational factors could significantly influence human performance.

World War II marked the development of new and complex machines and weaponry, and these made new demands on operators' cognition. It was no longer possible to adopt the Tayloristic principle of matching individuals to preexisting jobs. Now the design of equipment had to take into account human limitations and take advantage of human capabilities. The decision-making, attention, situational awareness and hand-eye coordination of the machine's operator became key in the success or failure of a task. There was substantial research conducted to determine the human capabilities and limitations that had to be accomplished. A lot of this research took off where the aeromedical research between the wars had left off. An example of this is the study done by Fitts and Jones (1947), who studied the most effective configuration of control knobs to be used in aircraft cockpits.

Much of this research transcended into other equipment with the aim of making the controls and displays easier for the operators to use. The entry of the terms "human factors" and "ergonomics" into the modern lexicon date from this period. It was observed that fully functional aircraft flown by the best-trained pilots, still crashed. In 1943 Alphonse Chapanis, a lieutenant in the U.S. Army, showed that this so-called "pilot error" could be greatly reduced when more logical and differentiable controls replaced confusing designs in airplane cockpits. After the war, the Army Air Force published 19 volumes summarizing what had been established from research during the war.

In the decades since World War II, HF&E has continued to flourish and diversify. Work by Elias Porter and others within the RAND Corporation after WWII extended the conception of HF&E. "As the thinking progressed, a new concept developed—that it was possible to view an organization such as an air-defense, man-machine system as a single organism and that it was possible to study the behavior of such an organism. It was the climate for a breakthrough." In the initial 20 years after the World War II, most activities were done by the "founding fathers": Alphonse Chapanis, Paul Fitts, and Small.

During the Cold War

The beginning of the Cold War led to a major expansion of Defense supported research laboratories. Also, many labs established during WWII started expanding. Most of the research following the war was military-sponsored. Large sums of money were granted to universities to conduct research. The scope of the research also broadened from small equipments to entire workstations and systems. Concurrently, a lot of opportunities started opening up in the civilian industry. The focus shifted from research to participation through advice to engineers in the design of equipment. After 1965, the period saw a maturation of the discipline. The field has expanded with the development of the computer and computer applications.

The Space Age created new human factors issues such as weightlessness and extreme g-forces. Tolerance of the harsh environment of space and its effects on the mind and body were widely studied

Information Age

The dawn of the Information Age has resulted in the related field of human–computer interaction (HCI). Likewise, the growing demand for and competition among consumer goods and electronics has resulted in more companies and industries including human factors in their product design. Using advanced technologies in human kinetics, body-mapping, movement patterns and heat zones, companies are able to manufacture purpose-specific garments, including full body suits, jerseys, shorts, shoes, and even underwear.

HF&E Organizations

Formed in 1946 in the UK, the oldest professional body for human factors specialists and ergonomists is The Chartered Institute of Ergonomics and Human Factors, formally known as *The Ergonomics Society*.

The Human Factors and Ergonomics Society (HFES) was founded in 1957. The Society's mission is to promote the discovery and exchange of knowledge concerning the characteristics of human beings that are applicable to the design of systems and devices of all kinds.

The International Ergonomics Association (IEA) is a federation of ergonomics and human factors societies from around the world. The mission of the IEA is to elaborate and advance ergonomics science and practice, and to improve the quality of life by expanding its scope of application and contribution to society. As of September 2008, the International Ergonomics Association has 46 federated societies and 2 affiliated societies.

Related Organizations

The Institute of Occupational Medicine (IOM) was founded by the coal industry in 1969. From the outset the IOM employed an ergonomics staff to apply ergonomics principles to the design of mining machinery and environments. To this day, the IOM continues ergonomics activities, especially in the fields of musculoskeletal disorders; heat stress and the ergonomics of personal protective equipment (PPE). Like many in occupational ergonomics, the demands and requirements of an ageing UK workforce are a growing concern and interest to IOM ergonomists.

The International Society of Automotive Engineers (SAE) is a professional organization for mobility engineering professionals in the aerospace, automotive, and commercial vehicle industries. The Society is a standards development organization for the engineering of powered vehicles of all kinds, including cars, trucks, boats, aircraft, and others. The Society of Automotive Engineers has established a number of standards used in the automotive industry and elsewhere. It encourages the design of vehicles in accordance with established Human Factors principles. It is one of the most influential organizations with respect to Ergonomics work in Automotive design. This society regularly holds conferences which address topics spanning all aspects of Human Factors/ Ergonomics.

Practitioners

Human factors practitioners come from a variety of backgrounds, though predominantly they are psychologists (from the various subfields of industrial and organizational psychology, engineering psychology, cognitive psychology, perceptual psychology, applied psychology, and experimental psychology) and physiologists. Designers (industrial, interaction, and graphic), anthropologists, technical communication scholars and computer scientists also contribute. Typically, an ergonomist will have an undergraduate degree in psychology, engineering, design or health sciences, and usually a masters degree or doctoral degree in a related discipline. Though some practitioners enter the field of human factors from other disciplines, both M.S. and PhD degrees in Human Factors Engineering are available from several universities worldwide.

Related Software

- ErgoFellow

- 3DSSPP

Methods

Until recently, methods used to evaluate human factors and ergonomics ranged from simple questionnaires to more complex and expensive usability labs. Some of the more common HF&E methods are listed below:

- Ethnographic analysis: Using methods derived from ethnography, this process focuses on observing the uses of technology in a practical environment. It is a qualitative and observational method that focuses on "real-world" experience and pressures, and the usage of technology or environments in the workplace. The process is best used early in the design process.

- Focus Groups are another form of qualitative research in which one individual will facilitate discussion and elicit opinions about the technology or process under investigation. This can be on a one to one interview basis, or in a group session. Can be used to gain a large quantity of deep qualitative data, though due to the small sample size, can be subject to a higher degree of individual bias. Can be used at any point in the design process, as it is largely dependent on the exact questions to be pursued, and the structure of the group. Can be extremely costly.

- Iterative design: Also known as prototyping, the iterative design process seeks to involve users at several stages of design, in order to correct problems as they emerge. As prototypes emerge from the design process, these are subjected to other forms of analysis as outlined in this article, and the results are then taken and incorporated into the new design. Trends amongst users are analyzed, and products redesigned. This can become a costly process, and needs to be done as soon as possible in the design process before designs become too concrete.

- Meta-analysis: A supplementary technique used to examine a wide body of already existing data or literature in order to derive trends or form hypotheses in order to aid design decisions. As part of a literature survey, a meta-analysis can be performed in order to discern a collective trend from individual variables.

- Subjects-in-tandem: Two subjects are asked to work concurrently on a series of tasks while vocalizing their analytical observations. The technique is also known as "Co-Discovery" as participants tend to feed off of each other's comments to generate a richer set of observations than is often possible with the participants separately. This is observed by the researcher, and can be used to discover usability difficulties. This process is usually recorded.

- Surveys and Questionnaires: A commonly used technique outside of Human Factors as well, surveys and questionnaires have an advantage in that they can be administered to a large group of people for relatively low cost, enabling the researcher to gain a large amount of data. The validity of the data obtained is, however, always in question, as the questions must be written and interpreted correctly, and are, by definition, subjective. Those who actually respond are in effect self-selecting as well, widening the gap between the sample and the population further.

- Task analysis: A process with roots in activity theory, task analysis is a way of systematically describing human interaction with a system or process to understand how to match the demands of the system or process to human capabilities. The complexity of this process is generally proportional to the complexity of the task being analyzed, and so can vary in cost and time involvement. It is a qualitative and observational process. Best used early in the design process.

- Think aloud protocol: Also known as "concurrent verbal protocol", this is the process of asking a user to execute a series of tasks or use technology, while continuously verbalizing their thoughts so that a researcher can gain insights as to the users' analytical process. Can be useful for finding design flaws that do not affect task performance, but may have a negative cognitive affect on the user. Also useful for utilizing experts in order to better understand procedural knowledge of the task in question. Less expensive than focus groups, but tends to be more specific and subjective.

- User analysis: This process is based around designing for the attributes of the intended user or operator, establishing the characteristics that define them, creating a persona for the user. Best done at the outset of the design process, a user analysis will attempt to predict the most common users, and the characteristics that they would be assumed to have in common. This can be problematic if the design concept does not match the actual user, or if the identified are too vague to make clear design decisions from. This process is, however, usually quite inexpensive, and commonly used.

- "Wizard of Oz": This is a comparatively uncommon technique but has seen some use in mobile devices. Based upon the Wizard of Oz experiment, this technique involves an operator who remotely controls the operation of a device in order to imitate the response of an actual computer program. It has the advantage of producing a highly changeable set of reactions, but can be quite costly and difficult to undertake.

- Methods Analysis is the process of studying the tasks a worker completes using a step-by-step investigation. Each task in broken down into smaller steps until each motion the worker performs is described. Doing so enables you to see exactly where repetitive or straining tasks occur.

- Time studies determine the time required for a worker to complete each task. Time studies are often used to analyze cyclical jobs. They are considered "event based" studies because time measurements are triggered by the occurrence of predetermined events.

- Work sampling is a method in which the job is sampled at random intervals to determine the proportion of total time spent on a particular task. It provides insight into how often workers are performing tasks which might cause strain on their bodies.

- Predetermined time systems are methods for analyzing the time spent by workers on a particular task. One of the most widely used predetermined time system is called Methods-Time-Measurement (MTM). Other common work measurement systems include MODAPTS and MOST. Industry specific applications based on PTS are Seweasy,MODAPTS and GSD as seen in paper: Miller, Doug, Towards Sustainable Labour Costing in UK Fashion Retail (February 5, 2013).

- Cognitive Walkthrough: This method is a usability inspection method in which the evaluators can apply user perspective to task scenarios to identify design problems. As applied to macroergonomics, evaluators are able to analyze the usability of work system designs to identify how well a work system is organized and how well the workflow is integrated.

- Kansei Method: This is a method that transforms consumer's responses to new products into design specifications. As applied to macroergonomics, this method can translate employee's responses to changes to a work system into design specifications.

- High Integration of Technology, Organization, and People (HITOP): This is a manual procedure done step-by-step to apply technological change to the workplace. It allows managers to be more aware of the human and organizational aspects of their technology plans, allowing them to efficiently integrate technology in these contexts.

- Top Modeler: This model helps manufacturing companies identify the organizational changes needed when new technologies are being considered for their process.

- Computer-integrated Manufacturing, Organization, and People System Design (CIMOP): This model allows for evaluating computer-integrated manufacturing, organization, and people system design based on knowledge of the system.

- Anthropotechnology: This method considers analysis and design modification of systems for the efficient transfer of technology from one culture to another.

- Systems Analysis Tool (SAT): This is a method to conduct systematic trade-off evaluations of work-system intervention alternatives.

- Macroergonomic Analysis of Structure (MAS): This method analyzes the structure of work systems according to their compatibility with unique sociotechnical aspects.

- Macroergonomic Analysis and Design (MEAD): This method assesses work-system processes by using a ten-step process.

- Virtual Manufacturing and Response Surface Methodology (VMRSM): This method uses computerized tools and statistical analysis for workstation design.

Weaknesses of HF&E Methods

Problems related to measures of usability include the fact that measures of learning and retention of how to use an interface are rarely employed and some studies treat measures of how users interact with interfaces as synonymous with quality-in-use, despite an unclear relation.

Although field methods can be extremely useful because they are conducted in the users' natural environment, they have some major limitations to consider. The limitations include:

1. Usually take more time and resources than other methods

2. Very high effort in planning, recruiting, and executing than other methods

3. Much longer study periods and therefore requires much goodwill among the participants

4. Studies are longitudinal in nature, therefore, attrition can become a problem.

Maintenance Engineering

Maintenance Engineering is the discipline and profession of applying engineering concepts to the optimization of equipment, procedures, and departmental budgets to achieve better maintainability, reliability, and availability of equipment.

Maintenance, and hence maintenance engineering, is increasing in importance due to rising amounts of equipment, systems, machineries and infrastructure. Since the Industrial Revolution, devices, equipment, machinery and structures have grown increasingly complex, requiring a host of personnel, vocations and related systems needed to maintain them. Prior to 2006, the United States spent approximately US$300 billion annually on plant maintenance and operations alone. Maintenance is to ensure a unit is fit for purpose, with maximum availability at minimum costs. A person practicing Maintenance Engineering is known as a Maintenance Engineer.

Maintenance Engineer's Essential Knowledge

A Maintenance Engineer should possess significant knowledge of statistics, probability and logistics, and additionally in the fundamentals of the operation of the equipment and machinery he or she is responsible for.

A Maintenance Engineer shall also possess high interpersonal, communication, management skills and ability to make quick decisions.

Typical Maintenance Engineering Responsibilities

Typical responsibilities include:

- Assure optimization of the Maintenance Organization structure
- Analysis of repetitive equipment failures
- Estimation of maintenance costs and evaluation of alternatives
- Forecasting of spare parts
- Assessing the needs for equipment replacements and establish replacement programs when due
- Application of scheduling and project management principles to replacement programs
- Assessing required maintenance tools and skills required for efficient maintenance of equipment
- Assessing required skills required for maintenance personnel
- Reviewing personnel transfers 2 and from maintenance organizations
- Assessing and reporting safety hazards associated with maintenance of equipment

Maintenance Engineering Education

Institutions across the world have recognised the need for Maintenance Engineering. Maintenance Engineers usually hold a degree in Mechanical Engineering, Industrial Engineering, or other Engineering Disciplines. In recent years specialised bachelor and master courses have developed. The Bachelor Degree program in Maintenance Engineering,at the German-Jordanian University in Amman is addressing the need, as well as the Bachelor Programme in Maintenance Engineering at Luleå University of Technology. With an increased demand for Chartered Engineers, The

University of Central Lancashire in United Kingdom has developed a MSc in Maintenance Engineering currently under accreditation with the Institution of Engineering and Technology and a Top-up Bachelor of Engineering with honour degree for technicians holding a Higher National Diploma and seeking a progression in their professional career.

Product Lifecycle

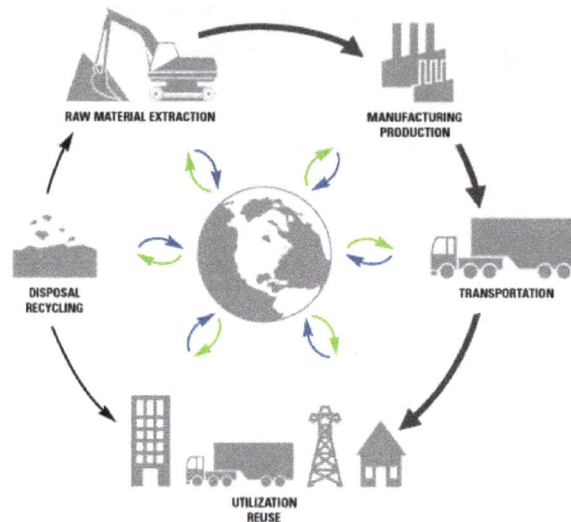

A generic lifecycle of products

In industry, product lifecycle management (PLM) is the process of managing the entire lifecycle of a product from inception, through engineering design and manufacture, to service and disposal of manufactured products. PLM integrates people, data, processes and business systems and provides a product information backbone for companies and their extended enterprise.

History

The inspiration for the burgeoning business process now known as PLM came from American Motors Corporation (AMC). The automaker was looking for a way to speed up its product development process to compete better against its larger competitors in 1985, according to François Castaing, Vice President for Product Engineering and Development. After introducing its compact Jeep Cherokee (XJ), the vehicle that launched the modern sport utility vehicle (SUV) market, AMC began development of a new model, that later came out as the Jeep Grand Cherokee. The first part in its quest for faster product development was computer-aided design (CAD) software system that make engineers more productive. The second part in this effort was the new communication system that allowed conflicts to be resolved faster, as well as reducing costly engineering changes because all drawings and documents were in a central database. The product data management was so effective that after AMC was purchased by Chrysler, the system was expanded throughout the enterprise connecting everyone involved in designing and building products. While an early adopter of PLM technology, Chrysler was able to become the auto industry's lowest-cost producer, recording development costs that were half of the industry average by the mid-1990s.

1982 - 1983 Rockwell Int'l developed initial concepts of PDM and PLM for the B-1B bomber program. A system called Engineering Data System (EDS) was augmented to interface with Computervision and CADAM systems to track part configurations and lifecycle of components and assemblies. A white paper on this topic was presented during those years at a Computervision User's Group meeting in San Diego. Shortly after Computervison released its system implementing only the PDM aspects as the lifecycle model was specific to Rockwell and Aerospace needs.

Forms

PLM systems help organizations in coping with the increasing complexity and engineering challenges of developing new products for the global competitive markets.

Product lifecycle management (PLM) should be distinguished from 'product life-cycle management (marketing)' (PLCM). PLM describes the engineering aspect of a product, from managing descriptions and properties of a product through its development and useful life; whereas, PLCM refers to the commercial management of life of a product in the business market with respect to costs and sales measures.

Product lifecycle management can be considered one of the four cornerstones of a manufacturing corporation's information technology structure. All companies need to manage communications and information with their customers (CRM-customer relationship management), their suppliers and fulfillment (SCM-supply chain), their resources within the enterprise (ERP-enterprise resource planning) and their product planning and development (PLM).

One form of PLM is called people-centric PLM. While traditional PLM tools have been deployed only on release or during the release phase, people-centric PLM targets the design phase.

As of 2009, ICT development (EU-funded PROMISE project 2004–2008) has allowed PLM to extend beyond traditional PLM and integrate sensor data and real time 'lifecycle event data' into PLM, as well as allowing this information to be made available to different players in the total lifecycle of an individual product (closing the information loop). This has resulted in the extension of PLM into closed-loop lifecycle management (CL_2M).

Benefits

Documented benefits of product lifecycle management include:

- Reduced time to market
- Increase full price sales
- Improved product quality and reliability
- Reduced prototyping costs

- More accurate and timely request for quote generation

- Ability to quickly identify potential sales opportunities and revenue contributions

- Savings through the re-use of original data

- A framework for product optimization

- Reduced waste

- Savings through the complete integration of engineering workflows

- Documentation that can assist in proving compliance for RoHS or Title 21 CFR Part 11

- Ability to provide contract manufacturers with access to a centralized product record

- Seasonal fluctuation management

- Improved forecasting to reduce material costs

- Maximize supply chain collaboration

Areas of PLM

Within PLM there are five primary areas;

1. Systems engineering (SE)

2. Product and portfolio m² (PPM)

3. Product design (CAx)

4. Manufacturing process management (MPM)

5. Product data management (PDM)

Note: While application software is not required for PLM processes, the business complexity and rate of change requires organizations execute as rapidly as possible.

Systems engineering is focused on meeting all requirements, primarily meeting customer needs, and coordinating the systems design process by involving all relevant disciplines. An important aspect for life cycle management is a subset within Systems Engineering called Reliability Engineering. Product and portfolio management is focused on managing resource allocation, tracking progress vs. plan for new product development projects that are in process (or in a holding status). Portfolio management is a tool that assists management in tracking progress on new products and making trade-off decisions when allocating scarce resources.Product design is the process of creating a new product to be sold by a business to its customers. Manufacturing process management is a collection of technologies and methods used to define how products are to be manufactured. Product data management is focused on capturing and maintaining information on products and/or services through their development and useful life. Change management is an important part of PDM/PLM.

Introduction to Development Process

The core of PLM (product lifecycle management) is in the creation and central management of all product data and the technology used to access this information and knowledge. PLM as a discipline emerged from tools such as CAD, CAM and PDM, but can be viewed as the integration of these tools with methods, people and the processes through all stages of a product's life. It is not just about software technology but is also a business strategy.

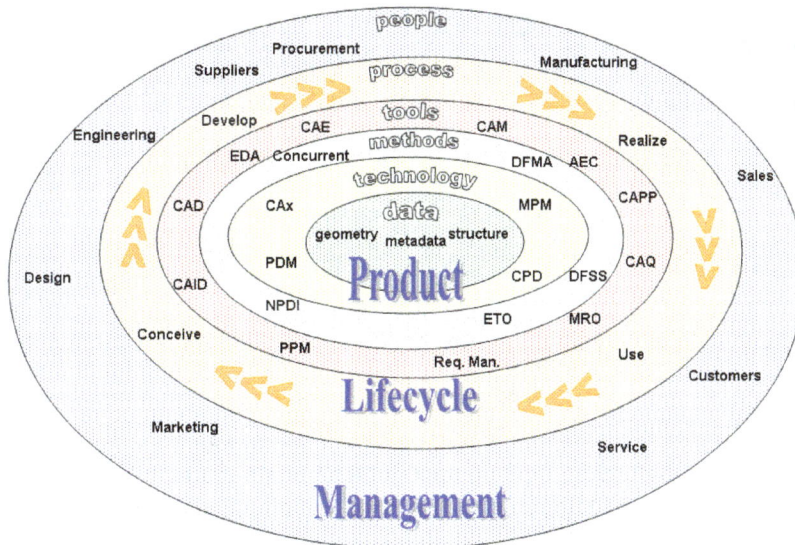

Product lifecycle management

For simplicity the stages described are shown in a traditional sequential engineering workflow. The exact order of event and tasks will vary according to the product and industry in question but the main processes are:

- Conceive
    - Specification
    - Concept design
- Design
    - Detailed design
    - Validation and analysis (simulation)
    - Tool design
- Realise
    - Plan manufacturing
    - Manufacture
    - Build/Assemble

- o   Test (quality control)
- Service
  - o   Sell and deliver
  - o   Use
  - o   Maintain and support
  - o   Dispose

The major key point events are:

- Order
- Idea
- Kickoff
- Design freeze
- Launch

The reality is however more complex, people and departments cannot perform their tasks in isolation and one activity cannot simply finish and the next activity start. Design is an iterative process, often designs need to be modified due to manufacturing constraints or conflicting requirements. Where a customer order fits into the time line depends on the industry type and whether the products are for example, built to order, engineered to order, or assembled to order.

Phases of Product Lifecycle and Corresponding Technologies

Many software solutions have been developed to organize and integrate the different phases of a product's lifecycle. PLM should not be seen as a single software product but a collection of software tools and working methods integrated together to address either single stages of the lifecycle or connect different tasks or manage the whole process. Some software providers cover the whole PLM range while others single niche application. Some applications can span many fields of PLM with different modules within the same data model. An overview of the fields within PLM is covered here. It should be noted however that the simple classifications do not always fit exactly, many areas overlap and many software products cover more than one area or do not fit easily into one category. It should also not be forgotten that one of the main goals of PLM is to collect knowledge that can be reused for other projects and to coordinate simultaneous concurrent development of many products. It is about business processes, people and methods as much as software application solutions. Although PLM is mainly associated with engineering tasks it also involves marketing activities such as product portfolio management (PPM), particularly with regards to new product development (NPD). There are several life-cycle models in industry to consider, but most are rather similar. What follows below is one possible life-cycle model; while it emphasizes hardware-oriented products, similar phases would describe any form of product or service, including non-technical or software-based products:

Phase 1: Conceive

Imagine, Specify, Plan, Innovate

The first stage is the definition of the product requirements based on customer, company, market and regulatory bodies' viewpoints. From this specification, the product's major technical parameters can be defined. In parallel, the initial concept design work is performed defining the aesthetics of the product together with its main functional aspects. Many different media are used for these processes, from pencil and paper to clay models to 3D CAID computer-aided industrial design software.

In some concepts, the investment of resources into research or analysis-of-options may be included in the conception phase – e.g. bringing the technology to a level of maturity sufficient to move to the next phase. However, life-cycle engineering is iterative. It is always possible that something doesn't work well in any phase enough to back up into a prior phase – perhaps all the way back to conception or research. There are many examples to draw from.

Phase 2: Design

Describe, Define, Develop, Test, Analyze and Validate

This is where the detailed design and development of the product's form starts, progressing to prototype testing, through pilot release to full product launch. It can also involve redesign and ramp for improvement to existing products as well as planned obsolescence. The main tool used for design and development is CAD. This can be simple 2D drawing / drafting or 3D parametric feature based solid/surface modeling. Such software includes technology such as Hybrid Modeling, Reverse Engineering, KBE (knowledge-based engineering), NDT (Nondestructive testing), and Assembly construction.

This step covers many engineering disciplines including: mechanical, electrical, electronic, software (embedded), and domain-specific, such as architectural, aerospace, automotive, ... Along with the actual creation of geometry there is the analysis of the components and product assemblies. Simulation, validation and optimization tasks are carried out using CAE (computer-aided engineering) software either integrated in the CAD package or stand-alone. These are used to perform tasks such as:- Stress analysis, FEA (finite element analysis); kinematics; computational fluid dynamics (CFD); and mechanical event simulation (MES). CAQ (computer-aided quality) is used for tasks such as Dimensional tolerance (engineering) analysis. Another task performed at this stage is the sourcing of bought out components, possibly with the aid of procurement systems.

Phase 3: Realize

Manufacture, Make, Build, Procure, Produce, Sell and Deliver

Once the design of the product's components is complete the method of manufacturing is defined. This includes CAD tasks such as tool design; creation of CNC Machining instructions for the product's parts as well as tools to manufacture those parts, using integrated or separate CAM computer-aided manufacturing software. This will also involve analysis tools for process simulation for operations such as casting, molding, and die press forming. Once the manufacturing method has been identified CPM comes into play. This involves CAPE (computer-aided production engineering) or CAP/CAPP – (production planning) tools for carrying out factory, plant and facility layout and production simulation. For example: press-line simulation; and industrial ergonomics; as well as tool selection management. Once components are manufactured their geometrical form and size can be checked against the original CAD data with the use of computer-aided inspection equipment and software. Parallel to the engineering tasks, sales product configuration and marketing documentation work take place. This could include transferring engineering data (geometry and part list data) to a web based sales configurator and other desktop publishing systems.

Phase 4: Service

Use, Operate, Maintain, Support, Sustain, Phase-Out, Retire, Recycle and Disposal

The final phase of the lifecycle involves managing of in service information. Providing customers and service engineers with support information for repair and maintenance, as well as waste management/recycling information. This involves using tools such as Maintenance, Repair and Operations Management (MRO) software.

There is an end-of-life to every product. Whether it be disposal or destruction of material objects or information, this needs to be considered since it may not be free from ramifications.

All Phases: Product Lifecycle

Communicate, Manage and Collaborate

None of the above phases can be seen in isolation. In reality a project does not run sequentially or in isolation of other product development projects. Information is flowing between different people and systems. A major part of PLM is the co-ordination and management of product definition data. This includes managing engineering changes and release status of components; configuration product variations; document management; planning project resources and timescale and risk assessment.

For these tasks graphical, text and metadata such as product bills of materials (BOMs) needs to be managed. At the engineering departments level this is the domain of PDM – (product data management) software, at the corporate level EDM (enterprise data management) software, these two definitions tend to blur however but it is typical to see two or more data management systems within an organization. These systems are also linked to other corporate systems such as SCM, CRM, and ERP. Associated with these system are project management Systems for project/program planning.

This central role is covered by numerous collaborative product development tools which run throughout the whole lifecycle and across organizations. This requires many technology tools in the areas of conferencing, data sharing and data translation. The field being product visualization which includes technologies such as DMU (digital mock-up), immersive virtual digital prototyping (virtual reality), and photo-realistic imaging.

User Skills

The broad array of solutions that make up the tools used within a PLM solution-set (e.g., CAD, CAM, CAx...) were initially used by dedicated practitioners who invested time and effort to gain the required skills. Designers and engineers worked wonders with CAD systems, manufacturing engineers became highly skilled CAM users while analysts, administrators and managers fully mastered their support technologies. However, achieving the full advantages of PLM requires the participation of many people of various skills from throughout an extended enterprise, each requiring the ability to access and operate on the inputs and output of other participants.

Despite the increased ease of use of PLM tools, cross-training all personnel on the entire PLM tool-set has not proven to be practical. Now, however, advances are being made to address ease of use for all participants within the PLM arena. One such advance is the availability of "role" specific user interfaces. Through tailorable user interfaces (UIs), the commands that are presented to users are appropriate to their function and expertise.

These techniques include:-

- Concurrent engineering workflow

- Industrial design

- Bottom–up design

- Top–down design

- Both-ends-against-the-middle design

- Front-loading design workflow

- Design in context

- Modular design

- NPD new product development

- DFSS design for Six Sigma

- DFMA design for manufacture / assembly

- Digital simulation engineering

- Requirement-driven design

- Specification-managed validation

- Configuration management

Concurrent Engineering Workflow

Concurrent engineering (British English: simultaneous engineering) is a workflow that, instead of working sequentially through stages, carries out a number of tasks in parallel. For example:

starting tool design as soon as the detailed design has started, and before the detailed designs of the product are finished; or starting on detail design solid models before the concept design surfaces models are complete. Although this does not necessarily reduce the amount of manpower required for a project, as more changes are required due to the incomplete and changing information, it does drastically reduce lead times and thus time to market.

Feature-based CAD systems have for many years allowed the simultaneous work on 3D solid model and the 2D drawing by means of two separate files, with the drawing looking at the data in the model; when the model changes the drawing will associatively update. Some CAD packages also allow associative copying of geometry between files. This allows, for example, the copying of a part design into the files used by the tooling designer. The manufacturing engineer can then start work on tools before the final design freeze; when a design changes size or shape the tool geometry will then update. Concurrent engineering also has the added benefit of providing better and more immediate communication between departments, reducing the chance of costly, late design changes. It adopts a problem prevention method as compared to the problem solving and re-designing method of traditional sequential engineering.

Bottom–Up Design

Bottom–up design (CAD-centric) occurs where the definition of 3D models of a product starts with the construction of individual components. These are then virtually brought together in sub-assemblies of more than one level until the full product is digitally defined. This is sometimes known as the review structure showing what the product will look like. The BOM contains all of the physical (solid) components; it may (but not also) contain other items required for the final product BOM such as paint, glue, oil and other materials commonly described as 'bulk items'. Bulk items typically have mass and quantities but are not usually modelled with geometry.

Bottom–up design tends to focus on the capabilities of available real-world physical technology, implementing those solutions which this technology is most suited to. When these bottom–up solutions have real-world value, bottom–up design can be much more efficient than top–down design. The risk of bottom–up design is that it very efficiently provides solutions to low-value problems. The focus of bottom–up design is "what can we most efficiently do with this technology?" rather than the focus of top–down which is "What is the most valuable thing to do?"

Top–Down Design

Top–down design is focused on high-level functional requirements, with relatively less focus on existing implementation technology. A top level spec is decomposed into lower and lower level structures and specifications, until the physical implementation layer is reached. The risk of a top–down design is that it will not take advantage of the most efficient applications of current physical technology, especially with respect to hardware implementation. Top–down design sometimes results in excessive layers of lower-level abstraction and inefficient performance when the Top–down model has followed an abstraction path which does not efficiently fit available physical-level technology. The positive value of top–down design is that it preserves a focus on the optimum solution requirements.

A part-centric top–down design may eliminate some of the risks of top–down design. This starts with a layout model, often a simple 2D sketch defining basic sizes and some major defining parameters. Industrial design brings creative ideas to product development. Geometry from this is associatively copied down to the next level, which represents different subsystems of the product. The geometry in the sub-systems is then used to define more detail in levels below. Depending on the complexity of the product, a number of levels of this assembly are created until the basic definition of components can be identified, such as position and principal dimensions. This information is then associatively copied to component files. In these files the components are detailed; this is where the classic bottom–up assembly starts.

The top–down assembly is sometime known as a control structure. If a single file is used to define the layout and parameters for the review structure it is often known as a skeleton file.

Defense engineering traditionally develops the product structure from the top down. The system engineering process prescribes a functional decomposition of requirements and then physical allocation of product structure to the functions. This top down approach would normally have lower levels of the product structure developed from CAD data as a bottom–up structure or design.

Both-Ends-Against-the-Middle Design

Both-ends-against-the-middle (BEATM) design is a design process that endeavors to combine the best features of top–down design, and bottom–up design into one process. A BEATM design process flow may begin with an emergent technology which suggests solutions which may have value, or it may begin with a top–down view of an important problem which needs a solution. In either case the key attribute of BEATM design methodology is to immediately focus at both ends of the design process flow: a top–down view of the solution requirements, and a bottom–up view of the available technology which may offer promise of an efficient solution. The BEATM design process proceeds from both ends in search of an optimum merging somewhere between the top–down requirements, and bottom–up efficient implementation. In this fashion, BEATM has been shown to genuinely offer the best of both methodologies. Indeed some of the best success stories from either top–down or bottom–up have been successful because of an intuitive, yet unconscious use of the BEATM methodology. When employed consciously, BEATM offers even more powerful advantages.

Front Loading Design and Workflow

Front loading is taking top–down design to the next stage. The complete control structure and review structure, as well as downstream data such as drawings, tooling development and CAM models, are constructed before the product has been defined or a project kick-off has been authorized. These assemblies of files constitute a template from which a family of products can be constructed. When the decision has been made to go with a new product, the parameters of the product are entered into the template model and all the associated data is updated. Obviously predefined associative models will not be able to predict all possibilities and will require additional work. The main principle is that a lot of the experimental/investigative work has already been completed. A lot of knowledge is built into these templates to be reused on new products. This does require additional resources "up front" but can drastically reduce the time between project kick-off and launch. Such methods do however require organizational changes, as considerable engineering efforts are moved into "offline" development departments. It can be seen as an analogy to creating a concept car to test new technology for future products, but in this case the work is directly used for the next product generation.

Design in Context

Individual components cannot be constructed in isolation. CAD and CaiD models of components are designed within the context of part or all of the product being developed. This is achieved using assembly modelling techniques. Other components' geometry can be seen and referenced within the CAD tool being used. The other components within the sub-assembly may or may not have been constructed in the same system, their geometry being translated from other collaborative product development (CPD) formats. Some assembly checking such as DMU is also carried out using product visualization software.

Product and Process Lifecycle Management (Pplm)

Product and process lifecycle management (PPLM) is an alternate genre of PLM in which the process by which the product is made is just as important as the product itself. Typically, this is the life sciences and advanced specialty chemicals markets. The process behind the manufacture of a given compound is a key element of the regulatory filing for a new drug application. As such, PPLM seeks to manage information around the development of the process in a similar fashion that baseline PLM talks about managing information around development of the product.

One variant of PPLM implementations are Process Development Execution Systems (PDES). They typically implement the whole development cycle of high-tech manufacturing technology developments, from initial conception, through development and into manufacture. PDES integrate people with different backgrounds from potentially different legal entities, data, information and knowledge and business processes.

Market Size

Total spending on PLM software and services was estimated in 2006 to be above $30 billion a year. Market growth estimates are in the area of 10%. There are several PLM Vendors in the market but the primary players include Dassault Systemes, Siemens, Oracle and PTC.

Pyramid of Production Systems

According to Malakooti (2013), there are five long-term objectives that should be considered in production systems:

- cost which can be measured in terms of monetary units and usually consists of fixed and variable cost.

- Productivity which can be measured in terms of the number of products produced during a period of time.

- Quality which can be measured, for example, in terms of customers' satisfaction.

- Flexibility, for example, ability of the system to produce variety of products.

- Ecological Soundness which can be measured in terms of biological and environmental impacts of the production system.

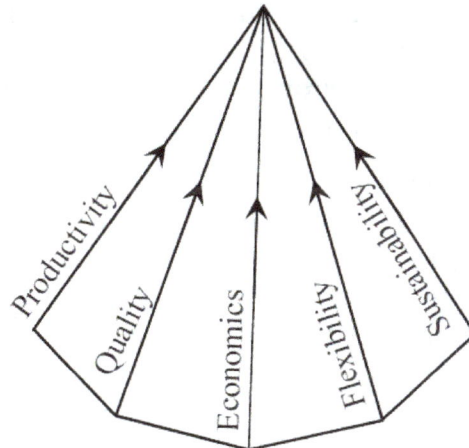

Pyramid of Production Systems

The relation between these five objects can be presented as pyramid which its tip is associated with the highest productivity, the highest quality, the most economical, the most flexibility, and the most sustainability. The points inside of this pyramid are associated with different combinations of five criteria. The tip of the pyramid is the ideal point but it is infeasible and the base of pyramid consists of the worst points.

Production Engineering

The Ford Motor Company's factory at Willow Run utilised Production Engineering principles to achieve record mass production of the B-24 Liberator military aircraft during World War II.

Production Engineering is a combination of manufacturing technology with management science. A production engineer typically has a wide knowledge of engineering practices and is aware of the management challenges related to production. The goal is to accomplish the production process in the smoothest, most-judicious and most-economic way.

Production Engineering encompasses the application of castings,machining processing, joining processes, metal cutting & tool design, metrology, machine tools, machining systems, automation, jigs and fixtures, die and mould design, material science, design of automobile parts, and machine designing and manufacturing. Production engineering also overlaps substantially with manufacturing engineering and industrial engineering.

In industry, once the design is realized, production engineering concepts regarding work-study, ergonomics, operation research, manufacturing management, materials management, production planning, etc., play important roles in efficient production processes. These deal with integrated design and efficient planning of the entire manufacturing system, which is becoming increasingly complex with the emergence of sophisticated production methods and control systems.

Production Engineer

The production engineer possesses a wide set of skills, competences and attitudes based on market and scientific knowledge. These abilities are fundamental for the performance of coordinating and integrating professionals of multidisciplinary teams. The production engineer should be able to:

- Scale and integrate resources. Usually required to consider physical, human and financial resources at high efficiency and low cost, yet considering the possibility of continuous further improvement;

- Make proper use of math and statistics to model production systems during decision making process;

- Design, implement and refine products, services, processes and systems taking in consideration that constraints and particularities of the related communities;

- Predict and analyze the demand. Select among scientific and technological appropriate knowledge in order to design, redesign or improve product/service functionality;

- Incorporate concepts and quality techniques along all the productive system. Deploy organizational standards for control proceedings and auditing;

- Stay up-to-date with technological developments, enabling them to enterprises and society;

- Understand the relation between production systems and the environment. This relates to the use of scarce resources, production rejects and sustainability;

- Manage and optimize flow (information and production flow).

Work opportunities are available in public and private sector manufacturing organizations engaged in implementation, development and management of new production processes, information and control systems, and computer controlled inspection, assembly and handling.

Packaging Engineering

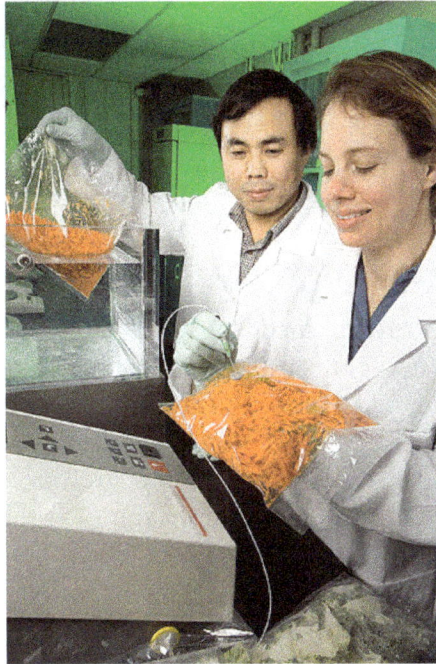

Testing modified atmosphere in a plastic bag of carrots

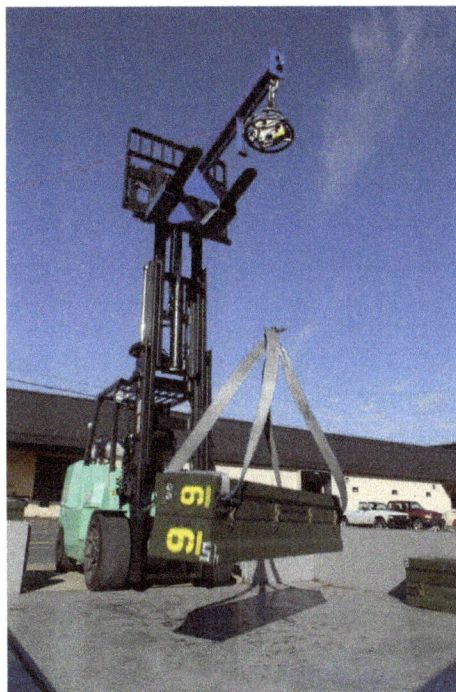

Military shipping container being drop tested

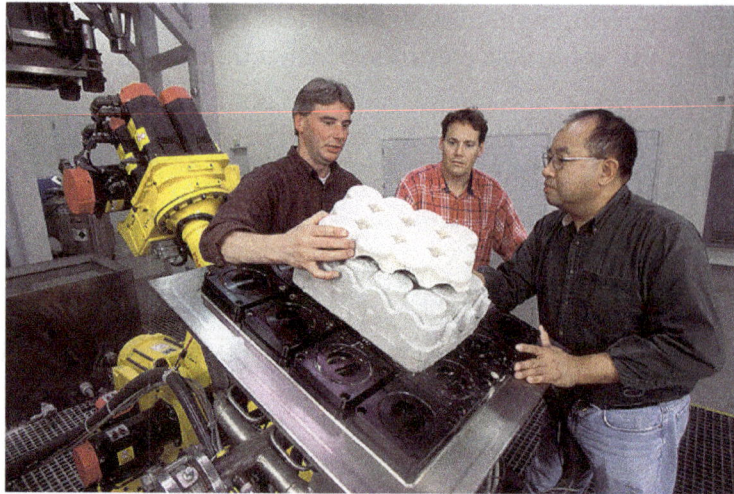

Engineers developing methods of molding packaging components from renewable resources such as straw

Packaging engineering, also package engineering, packaging technology and packaging science, is a broad topic ranging from design conceptualization to product placement. All steps along the manufacturing process, and more, must be taken into account in the design of the package for any given product. Package engineering includes industry-specific aspects of industrial engineering, marketing, materials science, industrial design and logistics. Packaging engineers must interact with research and development, manufacturing, marketing, graphic design, regulatory, purchasing, planning and so on. The package must sell and protect the product, while maintaining an efficient, cost-effective process cycle.

Engineers develop packages from a wide variety of rigid and flexible materials. Some materials have scores or creases to allow controlled folding into package shapes (sometimes resembling origami). Packaging involves extrusion, thermoforming, molding and other processing technologies. Packages are often developed for high speed fabrication, filling, processing, and shipment. Packaging engineers use principles of structural analysis and thermal analysis in their evaluations.

Education

Some packaging engineers have backgrounds in other science, engineering, or design disciplines while some have college degrees specializing in this field.

Formal packaging programs might be listed as package engineering, packaging science, packaging technology, etc. BE, BS, MS, M.Tech and PhD programs are available. Students in a packaging program typically begin with generalized science, business, and engineering classes before progressing into industry-specific topics such as shelf life stability, corrugated box design, cushioning, engineering design, labeling regulations, project management, food safety, robotics, RFID tags, quality management, package testing, packaging machinery, tamper-evident methods, recycling, computer-aided design, etc.

History

In 1952, Michigan State University became the first university in the USA to offer a degree in Packaging. Faculty and staff are widely published in packaging and other technical subjects.

The University of Wisconsin-Stout is one of only a few schools in the United States (and the only school in the UW System) that offer a B.S. degree in Packaging. The program combines general education with technical and professional studies. The packaging program involves the use of materials, methods, design concepts and machinery to develop and produce the packages that protect and preserve a product, help market the product and instruct the consumer in its proper use. The UW-Stout program offers the opportunity to apply theory to real problems. Through laboratory and co-op work experiences, you will apply the principles of science, mathematics and communications skills.

Throughout its history, UW-Stout has developed technical programs to meet the needs of business and industry. These programs have focused upon the direct application of technical knowledge to the solution of practical problems. The UW-Stout program leads to interesting, challenging, and rewarding careers in package printing; business and sales; foods and packaging; package graphic design; manufacturing and quality management; or package design, research and development.

Rutgers University offers the nation's only packaging program housed in an engineering school, since 1965. Unique from other engineering disciplines in its dynamic, the Packaging Engineering Concentration prepares engineering students by drawing heavily from each of the chemical, industrial, materials and mechanical engineering fields. It requires a strong background in mathematics, chemistry, physics, and computers. The Rutgers Packaging Engineering Program also offers an 18 credit Graduate Packaging Engineering Certificate.

Clemson University offers a degree in Packaging Science. There are also master's and doctorate programs. The undergraduate program at Clemson requires student's to take at least one six-month co-op for the degree.

Rochester Institute of Technology's Packaging Science program was first established in 1972. It is an interdisciplinary degree that leads to either a bachelors or masters of science. RIT requires all Packaging Science students to complete a 6-month internship/Co-op.

in Iloilo City in the Philippines in 2006, is the first to offer (and considered as the first in Asia) in the Philippines a (5-6 year) bachelor's degree in Packaging Engineering. It hosts the Philippine Center for Packaging Engineering and Technology that provides packaging engineering services and consultancy. University alumnus Dr. Lejo C. Brana, considered the founder of the packaging engineering program of the University, is a United States Hall of Famer in Packaging Engineering.

Indian Institute of Technology (IIT)-Roorkee started Packaging Technology in 2015. It is an interdisciplinary degree that leads to bachelors of technology.

In the United Kingdom, the Packaging Society, formerly the Institute of Packaging, offers the industry-standard *Diploma in Packaging Technology*. In India, The Indian institute of Packaging established in 1966, a national apex body which was set up in 1966 by the packaging and allied industries and the Ministry of Commerce, Government of India offers the 2 years Postgraduate Diploma in packaging as well as Graduate Diploma and certificate courses.It has its centers in Mumbai, Kolkata, chennai, Hyderabad and Delhi.Government Polytechnic Nagpur an autonomous institute run by Maharashtra Government offers 3 years Diploma In Packaging Technology.

Occupational Safety and Health

This painting depicts a woman examining her work on a lathe at a factory in Britain during World War II. Her eyes are not protected. Today, such practice would not be permitted in most industrialized countries that adhere to occupational health and safety standards for workers. In many countries however, such standards are still either weak or nonexistent.

Occupational safety and health (OSH), also commonly referred to as occupational health and safety (OHS), occupational health, or workplace health and safety (WHS), is a multidisciplinary field concerned with the safety, health, and welfare of people at work. These terms of course also refer to the goals of this field, so their use in the sense of this article was originally an abbreviation of *occupational safety and health program/department* etc.

The goals of occupational safety and health programs include to foster a safe and healthy work environment. OSH may also protect co-workers, family members, employers, customers, and many others who might be affected by the workplace environment. In the United States, the term occupational health and safety is referred to as occupational health and occupational and non-occupational safety and includes safety for activities outside of work.

In common-law jurisdictions, employers have a common law duty to take reasonable care of the safety of their employees. Statute law may in addition impose other general duties, introduce specific duties, and create government bodies with powers to regulate workplace safety issues: details of this vary from jurisdiction to jurisdiction.

All organizations have the duty to ensure that employees and any other person who may be affected by the organization's activities remain safe at all times.

Definition

As defined by the World Health Organization (WHO) "occupational health deals with all aspects of health and safety in the workplace and has a strong focus on primary prevention of hazards." Health has been defined as "a state of complete physical, mental and social well-being and not merely the absence of disease or infirmity." Occupational health is a multidisciplinary field of healthcare concerned with enabling an individual to undertake their occupation, in the way that causes least harm to their health. Health has been defined as It contrasts, for example, with the promotion of health and safety at work, which is concerned with preventing harm from any incidental hazards, arising in the workplace.

Workers cutting marble without any protective gear, Indore, India

Since 1950, the International Labour Organization (ILO) and the World Health Organization (WHO) have shared a common definition of occupational health. It was adopted by the Joint ILO/WHO Committee on Occupational Health at its first session in 1950 and revised at its twelfth session in 1995. The definition reads:

"The main focus in occupational health is on three different objectives: (i) the maintenance and promotion of workers' health and working capacity; (ii) the improvement of working environment and work to become conducive to safety and health and (iii) development of work organizations and working cultures in a direction which supports health and safety at work and in doing so also promotes a positive social climate and smooth operation and may enhance productivity of the undertakings. The concept of working culture is intended in this context to mean a reflection of the essential value systems adopted by the undertaking concerned. Such a culture is reflected in practice in the managerial systems, personnel policy, principles for participation, training policies and quality management of the undertaking."

—Joint ILO/WHO Committee on Occupational Health

Those in the field of occupational health come from a wide range of disciplines and professions including medicine, psychology, epidemiology, physiotherapy and rehabilitation, occupational therapy, occupational medicine, human factors and ergonomics, and many others. Professionals advise on a broad range of occupational health matters. These include how to avoid particular pre-existing conditions causing a problem in the occupation, correct posture for the work, frequency of rest breaks, preventative action that can be undertaken, and so forth.

"Occupational health should aim at: the promotion and maintenance of the highest degree of physical, mental and social well-being of workers in all occupations; the prevention amongst workers of departures from health caused by their working conditions; the protection of workers in their employment from risks resulting from factors adverse to health; the placing and maintenance of the worker in an occupational environment adapted to his physiological and psychological capabilities; and, to summarize, the adaptation of work to man and of each man to his job.

History

Harry McShane, age 16, 1908. Pulled into machinery in a factory in Cincinnati and had his arm ripped off at the shoulder and his leg broken without any compensation.

The research and regulation of occupational safety and health are a relatively recent phenomenon. As labor movements arose in response to worker concerns in the wake of the industrial revolution, worker's health entered consideration as a labor-related issue.

In the United Kingdom, the Factory Acts of the early nineteenth century (from 1802 onwards) arose out of concerns about the poor health of children working in cotton mills: the Act of 1833 created a dedicated professional Factory Inspectorate. The initial remit of the Inspectorate was to police restrictions on the working hours in the textile industry of children and young persons (introduced to prevent chronic overwork, identified as leading directly to ill-health and deformation, and indirectly to a high accident rate). However, on the urging of the Factory Inspectorate, a further Act in 1844 giving similar restrictions on working hours for women in the textile industry introduced a requirement for machinery guarding (but only in the textile industry, and only in areas that might be accessed by women or children).

In 1840 a Royal Commission published its findings on the state of conditions for the workers of the mining industry that documented the appallingly dangerous environment that they had to work in and the high frequency of accidents. The commission sparked public outrage which resulted in the Mines Act of 1842. The act set up an inspectorate for mines and collieries which resulted in many prosecutions and safety improvements, and by 1850, inspectors were able to enter and inspect premises at their discretion.

inaugurated the first social insurance legislation in 1883 and the first worker's compensation law in 1884 – the first of their kind in the Western world. Similar acts followed in other countries, partly in response to labor unrest.

Workplace Hazards

Although work provides many economic and other benefits, a wide array of workplace hazards also present risks to the health and safety of people at work. These include but are not limited to, "chemicals, biological agents, physical factors, adverse ergonomic conditions, allergens, a complex network of safety risks," and a broad range of psychosocial risk factors. Personal protective equipment can help protect against many of these hazards.

Physical hazards affect many people in the workplace. Occupational hearing loss is the most common work-related injury in the United States, with 22 million workers exposed to hazardous noise levels at work and an estimated $242 million spent annually on worker's compensation for hearing loss disability. Falls are also a common cause of occupational injuries and fatalities, especially in construction, extraction, transportation, healthcare, and building cleaning and maintenance. Machines have moving parts, sharp edges, hot surfaces and other hazards with the potential to crush, burn, cut, shear, stab or otherwise strike or wound workers if used unsafely.

(biohazards) include infectious microorganisms such as viruses and toxins produced by those organisms such as anthrax. Biohazards affect workers in many industries; influenza, for example, affects a broad population of workers. Outdoor workers, including farmers, landscapers, and construction workers, risk exposure to numerous biohazards, including animal bites and stings, urushiol from poisonous plants, and diseases transmitted through animals such as the West Nile virus and Lyme disease. Health care workers, including veterinary health workers, risk exposure to blood-borne pathogens and various infectious diseases, especially those that are emerging.

Dangerous chemicals can pose a chemical hazard in the workplace. There are many classifications of hazardous chemicals, including neurotoxins, immune agents, dermatologic agents, carcinogens, reproductive toxins, systemic toxins, asthmagens, pneumoconiotic agents, and sensitizers. Authorities such as regulatory agencies set occupational exposure limits to mitigate the risk of chemical hazards. An international effort is investigating the health effects of mixtures of chemicals. There is some evidence that certain chemicals are harmful at lower levels when mixed with one or more other chemicals. This may be particularly important in causing cancer.

include risks to the mental and emotional well-being of workers, such as feelings of job insecurity, long work hours, and poor work-life balance. A recent Cochrane review - using moderate quality evidence - related that the addition of work-directed interventions for depressed workers receiving clinical interventions reduces the number of lost work days as compared to clinical interventions alone. This review also demonstrated that the addition of cognitive behavioral therapy to primary or occupational care and the addition of a "structured telephone outreach and care management program" to usual care are both effective at reducing sick leave days.

By Industry

Specific occupational safety and health risk factors vary depending on the specific sector and industry. Construction workers might be particularly at risk of falls, for instance, whereas fishermen might be particularly at risk of drowning. The United States Bureau of Labor Statistics identifies the fishing, aviation, lumber, metalworking, agriculture, mining and transportation industries as among some of the more dangerous for workers. Similarly psychosocial risks such as workplace violence are more pronounced for certain occupational groups such as health care employees, police, correctional officers and teachers.

Construction

Workplace safety notices at the entrance of a Chinese construction site.

Construction is one of the most dangerous occupations in the world, incurring more occupational fatalities than any other sector in both the United States and in the European Union. In 2009, the fatal occupational injury rate among construction workers in the United States was nearly three times that for all workers. Falls are one of the most common causes of fatal and non-fatal injuries among construction workers. Proper safety equipment such as harnesses and guardrails and procedures such as securing ladders and inspecting scaffolding can curtail the risk of occupational injuries in the construction industry. Due to the fact that accidents may have disastrous consequences for employees as well as organizations, it is of utmost importance to ensure health and safety of workers and compliance with HSE construction requirements. Health and safety legislation in the construction industry involves many rules and regulations. For example, the role of the Construction Design Management (CDM) Coordinator as a requirement has been aimed at improving health and safety on-site.

The 2010 National Health Interview Survey Occupational Health Supplement (NHIS-OHS) identified work organization factors and occupational psychosocial and chemical/physical exposures which may increase some health risks. Among all U.S. workers in the construction sector, 44% had non-standard work arrangements (were not regular permanent employees) compared to 19% of all U.S. workers, 15% had temporary employment compared to 7% of all U.S. workers, and 55% experienced job insecurity compared to 32% of all U.S. workers. Prevalence rates for exposure to physical/chemical hazards were especially high for the construction sector. Among nonsmoking workers, 24% of construction workers were exposed to secondhand smoke while only 10% of all

U.S. workers were exposed. Other physical/chemical hazards with high prevalence rates in the construction industry were frequently working outdoors (73%) and frequent exposure to vapors, gas, dust, or fumes (51%).

Agriculture

Rollover protection bar on a Fordson tractor.

Agriculture workers are often at risk of work-related injuries, lung disease, noise-induced hearing loss, skin disease, as well as certain cancers related to chemical use or prolonged sun exposure. On industrialized farms, injuries frequently involve the use of agricultural machinery. The most common cause of fatal agricultural injuries in the United States is tractor rollovers, which can be prevented by the use of roll over protection structures which limit the risk of injury in case a tractor rolls over. Pesticides and other chemicals used in farming can also be hazardous to worker health, and workers exposed to pesticides may experience illnesses or birth defects. As an industry in which families, including children, commonly work alongside their families, agriculture is a common source of occupational injuries and illnesses among younger workers. Common causes of fatal injuries among young farm worker include drowning, machinery and motor vehicle-related accidents.

The 2010 NHIS-OHS found elevated prevalence rates of several occupational exposures in the agriculture, forestry, and fishing sector which may negatively impact health. These workers often worked long hours. The prevalence rate of working more than 48 hours a week among workers employed in these industries was 37%, and 24% worked more than 60 hours a week. Of all workers in these industries, 85% frequently worked outdoors compared to 25% of all U.S. workers. Additionally, 53% were frequently exposed to vapors, gas, dust, or fumes, compared to 25% of all U.S. workers.

Service Sector

As the number of service sector jobs has risen in developed countries, more and more jobs have become sedentary, presenting a different array of health problems than those associated with manufacturing and the primary sector. Contemporary problems such as the growing rate of obesity and issues relating to occupational stress, workplace bullying, and overwork in many countries have further complicated the interaction between work and health.

According to data from the 2010 NHIS-OHS, hazardous physical/chemical exposures in the service sector were lower than national averages. On the other hand, potentially harmful work organization characteristics and psychosocial workplace exposures were relatively common in this sector. Among all workers in the service industry, 30% experienced job insecurity in 2010, 27% worked non-standard shifts (not a regular day shift), 21% had non-standard work arrangements (were not regular permanent employees).

Mining and Oil & Gas Extraction

According to data from the 2010 NHIS-OHS, workers employed in mining and oil & gas extraction industries had high prevalence rates of exposure to potentially harmful work organization characteristics and hazardous chemicals. Many of these workers worked long hours: 50% worked more than 48 hours a week and 25% worked more than 60 hours a week in 2010. Additionally, 42% worked non-standard shifts (not a regular day shift). These workers also had high prevalence of exposure to physical/chemical hazards. In 2010, 39% had frequent skin contact with chemicals. Among nonsmoking workers, 28% of those in mining and oil and gas extraction industries had frequent exposure to secondhand smoke at work. About two-thirds were frequently exposed to vapors, gas, dust, or fumes at work.

Healthcare and Social Assistance

Healthcare workers are exposed to many hazards that can adversely affect their health and well-being. Long hours, changing shifts, physically demanding tasks, violence, and exposures to infectious diseases and harmful chemicals are examples of hazards that put these workers at risk for illness and injury.

According to the Bureau of Labor statistics, U.S. hospitals recorded 253,700 work-related injuries and illnesses in 2011, which is 6.8 work-related injuries and illnesses for every 100 full-time employees. The injury and illness rate in hospitals is higher than the rates in construction and manufacturing – two industries that are traditionally thought to be relatively hazardous.

The Occupational Health Safety Network (OHSN) is a secure electronic surveillance system developed by the National Institute for Occupational Safety and Health (NIOSH) to address health and safety risks among health care personnel. OHSN uses existing data to characterize risk of injury and illness among health care workers. Hospitals and other healthcare facilities can upload the occupational injury data they already collect to the secure database for analysis and benchmarking with other de-identified facilities. NIOSH works with OHSN participants in identifying and implementing timely and targeted interventions. OHSN modules currently focus on three high risk and preventable events that can lead to injuries or musculoskeletal disorders among healthcare personnel: musculoskeletal injuries from patient handling activities; slips, trips, and falls; and workplace violence. OHSN enrollment is open to all healthcare facilities.

Workplace Fatality and Injury Statistics

United States

The Bureau of Labor Statistics of the United States Department of Labor compiles information about workplace fatalities and non-fatal injuries in the United States. In 1970, an estimated 14,000 workers were killed on the job – by 2010, the workforce had doubled, but workplace deaths were down to about 4,500. Between 1913 and 2013, workplace fatalities dropped by approximately 80%.

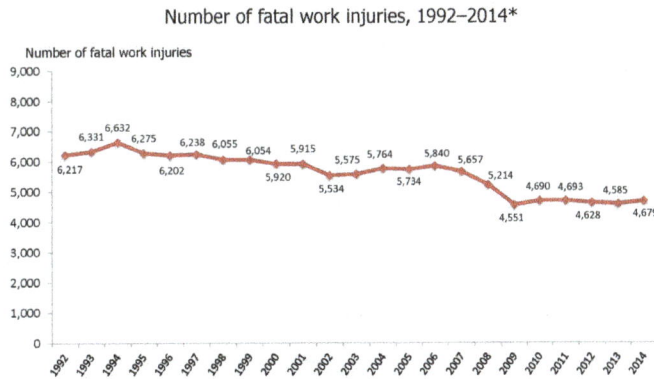

Number of occupational fatal work injuries in the U.S. from 1992 until 2014. Note, 2001 statistics do not include death related to the September 11 terrorist attacks.

The Bureau also compiles information about the most dangerous jobs. According to the census of occupational injuries 4,679 people died on the job in 2014.

Job	Fatalities	Fatalities per 100,000 employees
Fishermen	22	80.8
Pilots	81	63.2
Timber cutter	77	109.5
Structural metal workers	15	25.2
Waste collectors	27	35.8
Farmers and ranchers	263	26.0
Power-line workers	25	19.2
Construction workers and miners	130	17.9
Roofers	81	46.2
Truck drivers and other drivers	835	23.4
All occupations	4,679	3.3

2014 employer-reported injuries and illnesses		
Industry	Rate per 100 full-time employees	Number
Agriculture, forestry, fishing and hunting	5.5	52,400

2014 employer-reported injuries and illnesses		
Industry	Rate per 100 full-time employees	Number
Mining, quarrying, and oil and gas extraction	2.0	17,900
Construction (private)	3.6	200,900
Manufacturing	4.0	483,300
Wholesale trade	2.9	163,100
Retail trade	3.6	416,100
Transportation and warehousing (private)	4.8	201,500
Utilities (private)	2.4	13,400
Information	1.4	35,300
Finance and insurance	0.7	34,800
Real estate, rental, and leasing	2.9	51,100
Professional, scientific, and technical services	0.9	69,900
Management of companies and enterprises	1.0	21,100
Administrative and support services	2.4	109,300
Waste management and remediation services (private)	5.1	19,900
Educational services (private)	2.1	38,500
Health care and social assistance (private)	4.5	612,500
Leisure and hospitality	3.6	337,500
State government: Education	4.1	31,100
State government: Health care and social assistance	8.1	43,800
State government: Justice, public order, and safety activities	6.1	43,200
Local government: Construction	8.6	8,700
Local government: Transportation and warehousing	7.5	17,900
Local government: Utilities	5.4	12,200
Local government: Education	4.1	225,100
Local government: Health care and social assistance	5.6	41,000
Local government: Justice, public order, and safety activities	9.5	84,200
All industries including state and local government	3.4	3,675,800

Cause of injury and illness	2014 rate per 10,000 full-time employees
Contact with objects	23.8
Fall to lower level	5.4
Fall on same level	18.8
Slips or trips without fall	4.4
Over-exertion in lifting/lowering	11.0
Repetitive motion	2.7
Exposure to harmful substances or environments	4.3
Transportation incidents	5.8
Fires and explosions	0.2
Violence and other injuries by person or animal	6.8
Total	107.1

Musculoskeletal injuries accounted for 32% of all employer-reported injuries and illnesses in 2014.

European Union

In most countries males comprise the vast majority of workplace fatalities. In the EU as a whole, 94% of death were of males. In the UK the disparity was even greater with males comprising 97.4% of workplace deaths. In the UK there were 171 fatal injuries at work in financial year 2011-2, compared with 651 in calendar year 1974; the fatal injury rate declined over that period from 2.9 fatalities per 100,000 workers to 0.6 per 100,000 workers

Management Systems

International

In 2001, the International Labor Organization (ILO) published ILO-OSH 2001, also titled "Guidelines a on occupational safety and health management systems" to assist organizations with introducing OSH management systems. These guidelines encourage continual improvement in employee health and safety, achieved via a constant process of policy, organization, planning & implementation, evaluation, and action for improvement, all supported by constant auditing to determine the success of OSH actions.

The ILO management system was created to assist employers to keep pace with rapidly shifting and competitive industrial environments. The ILO recognizes that national legislation is essential, but sometimes insufficient on its own to address the challenges faced by industry, and therefore elected to ensure free and open distribution of administrative tools in the form of occupational health and safety management system guidance for everyone. This open access forum is intended to provide the tools for industry to create safe and healthy working environments and foster positive safety cultures within the organizations.

is an international occupational health and safety management system specification developed by the London-based BSI Group, a multinational business chiefly concerned with the production and distribution of standards related services. comprises two parts, OHSAS 18001 and 18002 and embraces a number of other publications. is the internationally recognized assessment specification for occupational health and safety management systems. It was developed by a selection of leading trade bodies, international standards and certification bodies to address a gap where no third-party certifiable international standard exists. This internationally recognized specification for occupational health and safety management system operates on the basis of policy, planning, implementation and operation, checking and corrective action, management review, and continual improvement.

The British Standards – Occupational Health and Safety management Systems Requirements Standard BS OHSAS 18001 was developed within the framework of the ISO standards series. Allowing it to integrate better into the larger system of ISO certifications. ISO 9001 Quality Management Systems and ISO 14001 Environmental Management System can work in tandem with BS OHSAS 18001/18002 to complement each other and form a better overall system. Each component of the system is specific, auditable, and accreditable by a third party after review.

Also Standards Australia and the Association Française de Normalisation (AFNOR) in France have developed occupational safety and health management standards.

United Kingdom

Guidance note HSG65: Successful Health and Safety Management, published by the British non-departmental public body Health and Safety Executive, was substantially re-written 2013. It now promotes the Plan Do Check Act approach to health and safety management, sharing similarities with BS OHSAS 18001. This achieved a balance between the original systems-based approach, and the more modern behavioural approach to safety management.

National Legislation and Public Organizations

Occupational safety and health practice vary among nations with different approaches to legislation, regulation, enforcement, and incentives for compliance. In the EU, for example, some member states promote OSH by providing public monies as subsidies, grants or financing, while others have created tax system incentives for OSH investments. A third group of EU member states has experimented with using workplace accident insurance premium discounts for companies or organisations with strong OSH records.

Australia

In Australia, the Commonwealth, four of the six states and both territories have enacted and administer harmonised Work Health and Safety Legislation in accordance with the *Intergovernmental Agreement for Regulatory and Operational Reform in Occupational Health and Safety*. Each of these jurisdictions has enacted Work Health & Safety legislation and regulations based on the Commonwealth Work Health and Safety Act 2011 and common Codes of Practice developed by Safe Work Australia. Some jurisdictions have also included mine safety under the model approach, however, most have retained separate legislation for the time being. Western Australia intends to adopt a moderated version of the model approach and Victoria has retained its own regime, although the Model WHS laws themselves drew heavily on the Victoria approach.

Canada

In Canada, workers are covered by provincial or federal labour codes depending on the sector in which they work. Workers covered by federal legislation (including those in mining, transportation, and federal employment) are covered by the Canada Labour Code; all other workers are covered by the health and safety legislation of the province in which they work. The Canadian Centre for Occupational Health and Safety (CCOHS), an agency of the Government of Canada, was created in 1966 by an Act of Parliament. The act was based on the belief that all Canadians had "...a fundamental right to a healthy and safe working environment." CCOHS is mandated to promote safe and healthy workplaces to help prevent work-related injuries and illnesses. The CCOHS maintains a useful (partial) list of OSH regulations for Canada and its provinces.

European Union

Number of full-time OSH inspectors per 100,000 full-time employees	
Italy	17.7
Finland	17.5
Denmark	11.9
United Kingdom	11.1
Norway	10.6
Sweden	10
Belgium	5.3
Netherlands	4.8
Ireland	4.5
Greece	4.1
France	3.5
Spain	2.1

In the European Union, member states have enforcing authorities to ensure that the basic legal requirements relating to occupational health and safety are met. In many EU countries, there is strong cooperation between employer and worker organisations (e.g. unions) to ensure good OSH performance as it is recognized this has benefits for both the worker (through maintenance of health) and the enterprise (through improved productivity and quality). In 1996, the European Agency for Safety and Health at Work was founded.

Member states of the European Union have all transposed into their national legislation a series of directives that establish minimum standards on occupational health and safety. These directives (of which there are about 20 on a variety of topics) follow a similar structure requiring the employer to assess the workplace risks and put in place preventive measures based on a hierarchy of control. This hierarchy starts with elimination of the hazard and ends with personal protective equipment.

However, certain EU member states admit to having lacking quality control in occupational safety services, to situations in which risk analysis takes place without any on-site workplace visits and to insufficient implementation of certain EU OSH directives. Based on this, it is hardly surprising that the total societal costs of work-related health problems and accidents vary from 2.6% to 3.8% of GNP between the EU member states.

Denmark

In Denmark, occupational safety and health is regulated by the Danish Act on Working Environment and cooperation at the workplace. The Danish Working Environment Authority carries out inspections of companies, draws up more detailed rules on health and safety at work and provides information on health and safety at work. The result of each inspection is made public on the web pages of the Danish Working Environment Authority so that the general public, current and prospective employees, customers and other stakeholders can inform themselves about whether a given organization has passed the inspection, should they wish to do so.

Spain

In Spain, occupational safety and health is regulated by the Spanish Act on Prevention of Labour Risks. The Ministry of Employment and Social Security is the authority responsible for issues relating to labour environment. The National Institute for Labour Safety and Hygiene is the technical public Organization specialized in occupational safety and health.

Sweden

In Sweden, occupational safety and health is regulated by the Work Environment Act. The Swedish Work Environment Authority is the government agency responsible for issues relating to the working environment. The agency should work to disseminate information and furnish advice on OSH, has a mandate to carry out inspections, and a right to issue stipulations and injunctions to any non-compliant employer.

United Kingdom

In the UK, health and safety legislation is drawn up and enforced by the Health and Safety Executive and local authorities (the local council) under the Health and Safety at Work etc. Act 1974 (HASAWA). HASAWA introduced (section 2) a general duty on an employer to ensure, so far as is reasonably practicable, the health, safety and welfare at work of all his employees; with the intention of giving a legal framework supporting 'codes of practice' not in themselves having legal force but establishing a strong presumption as to what was reasonably practicable (deviations from them could be justified by appropriate risk assessment). The previous reliance on detailed prescriptive rule-setting was seen as having failed to respond rapidly enough to technological change, leaving new technologies potentially un-regulated or inappropriately regulated. HSE has continued to make some regulations giving absolute duties (where something must be done with no 'reasonable practicability' test) but in the UK the regulatory trend is away from prescriptive rules, and towards 'goal setting' and risk assessment. Recent major changes to the laws governing asbestos and fire safety management embrace the concept of risk assessment. The other key aspect of the UK legislation is a statutory mechanism for worker involvement through elected health and safety representatives and health and safety committees. This followed a similar approach in Scandinavia, and that approach has since been adopted in Australia, Canada, New Zealand and Malaysia, for example.

For the UK, the government organisation dealing with occupational health has been the Employment Medical Advisory Service but in 2014 a new occupational health organisation - the Health and Work Service - was created to provide advice and assistance to employers in order to get back to work employees on long-term sick-leave. The service, funded by government, will offer medical assessments and treatment plans, on a voluntary basis, to people on long term absence from their employer; in return, the government will no longer foot the bill for Statutory Sick Pay provided by the employer to the individual.

India

In India, the Labour Ministry formulates national policies on occupational safety and health in factories and docks with advice and assistance from Directorate General of Factory Advice Service and Labour Institutes (DGFASLI), and enforces its Policies through inspectorates of factories and inspectorates of dock safety. DGFASLI is the technical arm of the Ministry of Labour & Employment, Government of India and advises the factories on various problems concerning safety, health, efficiency and well - being of the persons at work places. The DGFASLI provides technical support in formulating rules, conducting occupational safety surveys and also for conducting occupational safety training programs.

Malaysia

In Malaysia, the Department of Occupational Safety and Health (DOSH) under the Ministry of Human Resource is responsible to ensure that the safety, health and welfare of workers in both the public and private sector is upheld. DOSH is responsible to enforce the Factories and Machinery Act 1967 and the Occupational Safety and Health Act 1994.

People's Republic of China

Hardware stores in China specializing in safety equipment

In the People's Republic of China, the Ministry of Health is responsible for occupational disease prevention and the State Administration of Work Safety for safety issues at work. On the provincial and municipal level, there are Health Supervisions for occupational health and local bureaus of Work Safety for safety. The "Occupational Disease Control Act of PRC" came into force on May 1, 2002. and Work safety Act of PRC on November 1, 2002. The Occupational Disease Control Act is under revision. The prevention of occupational disease is still in its initial stage compared with industrialised countries such as the US or UK.

Singapore

In Singapore, the Ministry of Manpower operates various checks and campaigns against unsafe work practices, such as when working at height, operating cranes and in traffic management. Examples include Operation Cormorant and the Falls Prevention Campaign.

South Africa

In South Africa the Department of Labour is responsible for occupational health and safety inspection and enforcement in commerce and industry apart from mining and energy production, where the Department of Mineral Resources is responsible.

The main statutory legislation on Health and Safety in the jurisdiction of the Department of Labour is *Act No. 85 of 1993: Occupational Health and Safety Act as amended by Occupational Health and Safety Amendment Act, No. 181 Of 1993*.

Regulations to the OHS Act include:

- *General Administrative Regulations, 2003*

- *Certificate of Competency Regulations, 1990*

- *Construction Regulations, 2003*

- *Diving Regulations 2009*

- *Driven Machinery Regulations, 1988*

- *Environmental Regulations for Workplaces, 1987*

- *General Machinery regulations, 1988*

- *General Safety Regulations, 1986*

- *Noise induced hearing loss regulations, 2003*

- *Pressure Equipment Regulations, 2004*

Taiwan

In Taiwan, the Occupational Safety and Health Administration of the Ministry of Labor is in charge of occupational safety and health. The matter is governed under the Occupational Safety and Health Act.

2007,official release the document TOSHMS (Taiwan Occupational Safety and Health Management System),which defined the basic rule about occupational safety standard.

United States

In the United States, President Richard Nixon signed the Occupational Safety and Health Act into law on December 29, 1970. The act created the three agencies that administer it. They include the Occupational Safety and Health Administration, National Institute for Occupational Safety and Health, and the Occupational Safety and Health Review Commission. The act authorized the Occupational Safety and Health Administration (OSHA) to regulate private employers in the 50 states, the District of Columbia, and territories. The Act establishing it includes a general duty clause (29 U.S.C. § 654, 5(a)) requiring an employer to comply with the Act and regulations derived from it, and to provide employees with "employment and a place of employment which are free from recognized hazards that are causing or are likely to cause death or serious physical harm to his employees."

OSHA was established in 1971 under the Department of Labor. It has headquarters in Washington, DC and ten regional offices, further broken down into districts, each organized into three sections; compliance, training, and assistance. Its stated mission is *to assure safe and healthful working conditions for working men and women by setting and enforcing standards and by providing training, outreach, education and assistance.* The original plan was for OSHA to oversee 50 state plans with OSHA funding 50% of each plan. Unfortunately it has not worked out that way. There are currently 26 approved state plans (4 cover only public employees) and no other states want to participate. OSHA manages the plan in the states not participating.

OSHA develops safety standards in the Code of Federal Regulation and enforces those safety standards through compliance inspections conducted by Compliance Officers; enforcement resources are focussed on high-hazard industries. Worksites may apply to enter OSHA's Voluntary Protection Program (VPP); a successful application leads to an on-site inspection ; if this is passed the site gains VPP status and OSHA no longer inspect it annually nor (normally) visit it unless there is a fatal accident or an employee complaint until VPP revalidation (after 3–5 years)(VPP sites have injury and illness rates less than half the average for their industry).

It has 73 specialists in local offices to provide tailored information and training to employers and employees at little or no cost Similarly OSHA produces a range of publications, provides advice to employers and funds consultation services available for small businesses.

OSHA's Alliance Program enables groups committed to worker safety and health to work with it to develop compliance assistance tools and resources, share information with workers and employers, and educate them about their rights and responsibilities. OSHA also has a Strategic Partnership Program that zeros in on specific hazards or specific geographic areas. OSHA manages Susan B. Harwood grants to nonprofit companies to train workers and employers to recognize, avoid, and prevent safety and health hazards in the workplace. Grants focus on small business, hard-to-reach workers and high-hazard industries.

The National Institute of Occupational Safety and Health (NIOSH), created under the same act, works closely with OSHA and provides the research behind many of OSHA's regulations and standards.

Professional Roles and Responsibilities

The roles and responsibilities of OSH professionals vary regionally, but may include evaluating working environments, developing, endorsing and encouraging measures that might prevent injuries and illnesses, providing OSH information to employers, employees, and the public, providing medical examinations, and assessing the success of worker health programs.

Europe

In Norway, the main required tasks of an occupational health and safety practitioner include the following:

- Systematic evaluations of the working environment

- Endorsing preventative measures which eliminate causes of illnesses in the workplace

- Providing information on the subject of employees' health

- Providing information on occupational hygiene, ergonomics, and environmental and safety risks in the workplace

In the Netherlands, the required tasks for health and safety staff are only summarily defined and include the following:

- Providing voluntary medical examinations

- Providing a consulting room on the work environment to the workers

- Providing health assessments (if needed for the job concerned)

'The main influence of the Dutch law on the job of the safety professional is through the requirement on each employer to use the services of a certified working conditions service to advise them on health and safety'. A 'certified service' must employ sufficient numbers of four types of certified experts to cover the risks in the organisations which use the service:

- A safety professional

- An occupational hygienist

- An occupational physician

- A work and organisation specialist.

In 2004, 37% of health and safety practitioners in Norway and 14% in the Netherlands had an MSc; 44% had a BSc in Norway and 63% in the Netherlands; and 19% had training as an OSH technician in Norway and 23% in the Netherlands.

USA

The main tasks undertaken by the OHS practitioner in the USA include:

- Develop processes, procedures, criteria, requirements, and methods to attain the best possible management of the hazards and exposures that can cause injury to people, and damage property, or the environment;

- Apply good business practices and economic principles for efficient use of resources to add to the importance of the safety processes;

- Promote other members of the company to contribute by exchanging ideas and other different approaches to make sure that every one in the corporation possess OHS knowledge and have functional roles in the development and execution of safety procedures;

- Assess services, outcomes, methods, equipment, workstations, and procedures by using qualitative and quantitative methods to recognise the hazards and measure the related risks;

- Examine all possibilities, effectiveness, reliability, and expenditure to attain the best results for the company concerned

Leather craftsman gloves, safety goggles, and a properly fitted hardhat are crucial for proper safety in a construction environment.

Knowledge required by the OHS professional in USA include:

- Constitutional and case law controlling safety, health, and the environment

- Operational procedures to plan/develop safe work practices

- Safety, health and environmental sciences

- Design of hazard control systems (i.e. fall protection, scaffoldings)

- Design of recordkeeping systems that take collection into account, as well as storage, interpretation, and dissemination

- Mathematics and statistics

- Processes and systems for attaining safety through design

Some skills required by the OHS professional in the USA include (but are not limited to):

- Understanding and relating to systems, policies and rules

- Holding checks and having control methods for possible hazardous exposures

- Mathematical and statistical analysis

Examining Manufacturing Hazards

- Planning safe work practices for systems, facilities, and equipment

- Understanding and using safety, health, and environmental science information for the improvement of procedures

- Interpersonal communication skills

Differences Between Countries and Regions

Because different countries take different approaches to ensuring occupational safety and health, areas of OSH need and focus also vary between countries and regions. Similar to the findings of the ENHSPO survey conducted in Australia, the Institute of Occupational Medicine in the UK found that there is a need to put greater emphasis on work-related illness in the UK. In contrast, in Australia and the USA a major responsibility of the OHS professional is to keep company directors and managers aware of the issues that they face in regards to occupational health and safety principles and legislation. However, in some other areas of Europe, it is precisely this which has been lacking: "Nearly half of senior managers and company directors do not have an up-to-date understanding of their health and safety-related duties and responsibilities."

Identifying Safety and Health Hazards

Hazards, Risks, Outcomes

The terminology used in OSH varies between countries, but generally speaking:

- A hazard is something that can cause harm if not controlled.

- The outcome is the harm that results from an uncontrolled hazard.

- A risk is a combination of the probability that a particular outcome will occur and the severity of the harm involved.

"Hazard", "risk", and "outcome" are used in other fields to describe e.g. environmental damage, or damage to equipment. However, in the context of OSH, "harm" generally describes the direct or indirect degradation, temporary or permanent, of the physical, mental, or social well-being of workers. For example, repetitively carrying out manual handling of heavy objects is a hazard. The outcome could be a musculoskeletal disorder (MSD) or an acute back or joint injury. The risk can be expressed numerically (e.g. a 0.5 or 50/50 chance of the outcome occurring during a year), in relative terms (e.g. "high/medium/low"), or with a multi-dimensional classification scheme (e.g. situation-specific risks).

Hazard Identification

Hazard identification or assessment is an important step in the overall risk assessment and risk management process. It is where individual work hazards are identified, assessed and controlled/eliminated as close to source (location of the hazard) as reasonably as possible. As technology, resources, social expectation or regulatory requirements change, hazard analysis focuses controls more closely toward the source of the hazard. Thus hazard control is a dynamic program of prevention. Hazard-based programs also have the advantage of not assigning or implying there are "acceptable risks" in the workplace. A hazard-based program may not be able to eliminate all risks, but neither does it accept "satisfactory" – but still risky – outcomes. And as those who calculate and manage the risk are usually managers while those exposed to the risks are a different group, workers, a hazard-based approach can by-pass conflict inherent in a risk-based approach.

The information that needs to be gathered from sources should apply to the specific type of work from which the hazards can come from. As mentioned previously, examples of these sources include interviews with people who have worked in the field of the hazard, history and analysis of past incidents, and official reports of work and the hazards encountered. Of these, the personnel interviews may be the most critical in identifying undocumented practices, events, releases, hazards and other relevant information. Once the information is gathered from a collection of sources, it is recommended for these to be digitally archived (to allow for quick searching) and to have a physical set of the same information in order for it to be more accessible. One innovative way to display the complex historical hazard information is with a historical hazards identification map, which distills the hazard information into an easy to use graphical format.

Risk Assessment

Modern occupational safety and health legislation usually demands that a risk assessment be carried out prior to making an intervention. It should be kept in mind that risk management requires risk to be managed to a level which is as low as is reasonably practical.

This assessment should:

- Identify the hazards

- Identify all affected by the hazard and how

- Evaluate the risk

- Identify and prioritize appropriate control measures

The calculation of risk is based on the likelihood or probability of the harm being realized and the severity of the consequences. This can be expressed mathematically as a quantitative assessment (by assigning low, medium and high likelihood and severity with integers and multiplying them to obtain a risk factor), or qualitatively as a description of the circumstances by which the harm could arise.

The assessment should be recorded and reviewed periodically and whenever there is a significant change to work practices. The assessment should include practical recommendations to control the risk. Once recommended controls are implemented, the risk should be re-calculated to determine of it has been lowered to an acceptable level. Generally speaking, newly introduced controls should lower risk by one level, i.e., from high to medium or from medium to low.

Contemporary Developments

On an international scale, the World Health Organization (WHO) and the International Labour Organization (ILO) have begun focusing on labour environments in developing nations with projects such as Healthy Cities. Many of these developing countries are stuck in a situation in which their relative lack of resources to invest in OSH leads to increased costs due to work-related illnesses and accidents. As a 2007 Factsheet from the European Agency for Safety and Health at Work states: "Countries with less developed OSH systems spend a far higher percentage of GDP on work-related injury and illness – taking resources away from more productive activities . . . The ILO estimates that work-related illness and accidents cost up to 10% of GDP in Latin America, compared with just 2.6% to 3.8% in the EU." There is continued use of asbestos, a notorious hazard, in some developing countries. So asbestos-related disease is, sadly, expected to continue to be a significant problem well into the future.

Nanotechnology

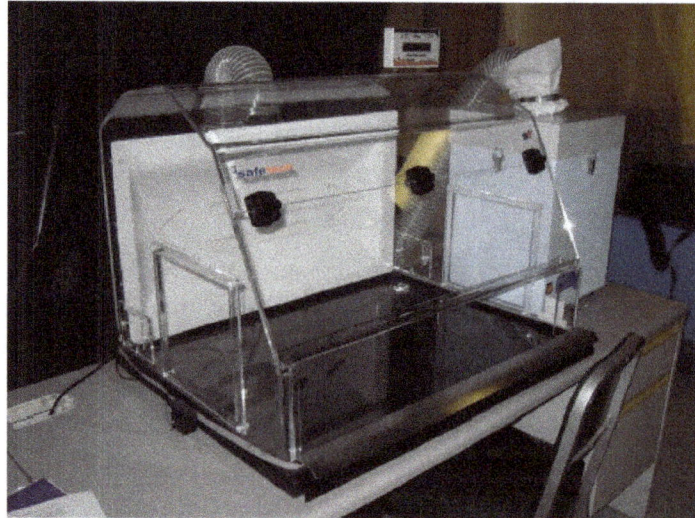

A nanomaterial containment hood, an example of an engineering control used to protect workers handling them on a regular basis.

Nanotechnology is an example of a new, relatively unstudied technology. A Swiss survey of one hundred thirty eight companies using or producing nanoparticulate matter in 2006 resulted in forty completed questionnaires. Sixty five per cent of respondent companies stated they did not have a formal risk assessment process for dealing with nanoparticulate matter. Nanotechnology already presents new issues for OSH professionals that will only become more difficult as nanostructures become more complex. The size of the particles renders most containment and personal protective equipment ineffective. The toxicology values for macro sized industrial substances are rendered inaccurate due to the unique nature of nanoparticulate matter. As nanoparticulate matter decreases in size its relative surface area increases dramatically, increasing any catalytic effect or chemical reactivity substantially versus the known value for the macro substance. This presents a new set of challenges in the near future to rethink contemporary measures to safeguard the health and welfare of employees against a nanoparticulate substance that most conventional controls have not been designed to manage.

Education

There are multiple levels of training applicable to the field of occupational safety and health (OSH). Programs range from individual non-credit certificates, focusing on specific areas of concern, to full doctoral programs. The University of Southern California was one of the first schools in the US to offer a Ph.D. program focusing on the field. Further, multiple master's degree programs exist, such as that of the Indiana State University who offer a master of science (MS) and a master of arts (MA) in OSH. Graduate programs are designed to train educators, as well as, high-level practitioners. Many OSH generalists focus on undergraduate studies; programs within schools, such as that of the University of North Carolina's online Bachelor of Science in Environmental Health and Safety, fill a large majority of hygienist needs. However, smaller companies often don't have full-time safety specialists on staff, thus, they appoint a current employee to the responsibility. Individuals finding themselves in positions such as these, or for those enhancing marketability in the job-search and promotion arena, may seek out a credit certificate program. For example, the University of Connecticut's online OSH Certificate, provides students familiarity with overarching concepts through a 15-credit (5-course) program. Programs such as these are often adequate tools in building a strong educational platform for new safety managers with a minimal outlay of time and money. Further, most hygienists seek certification by organizations which train in specific areas of concentration, focusing on isolated workplace hazards. The American Society for Safety Engineers (ASSE), American Board of Industrial Hygiene (ABIH), and American Industrial Hygiene Association (AIHA) offer individual certificates on many different subjects from forklift operation to waste disposal and are the chief facilitators of continuing education in the OSH sector. In the U.S. the training of safety professionals is supported by National Institute for Occupational Safety and Health through their NIOSH Education and Research Centers. In Australia, training in OSH is available at the vocational education and training level, and at university undergraduate and postgraduate level. Such university courses may be accredited by an Accreditation Board of the Safety Institute of Australia. The Institute has produced a Body of Knowledge which it considers is required by a generalist safety and health professional, and offers a professional qualification based on a four-step assessment.

World Day for Safety and Health at Work

On April 28 The International Labour Organisation celebrates "World Day for Safety and Health" to raise awareness of safety in the workplace. Occurring annually since 2003, each year it focuses on a specific area and bases a campaign around the theme.

Process Engineering

Process engineering focuses on the design, operation, control, and optimization of chemical, physical, and biological processes. Process engineering encompasses a vast range of industries, such as chemical, petrochemical, agriculture, mineral processing, advanced material, food, pharmaceutical, software development and biotechnological industries.

Process engineering involves translating the needs of the customer into (typically) production facilities that convert "raw materials" into value-added components that are transported to the next stage of the supply chain, typically "packaging engineering", but some larger volume processes such as petroleum refining tend to transfer the products into transportation (trucks or rail) that are then directed to distributors or bulk outlets. Prior to construction, the design work of process engineering begins with a "block diagram" showing raw materials and the transformations/unit operations desired. The design work then progresses to a Process flow diagram (PFD) where material flow paths, storage equipment (such as tanks and silos), transformations/Unit Operations (such as distillation columns, receiver/head tanks, mixing, separations, pumping, etc.) and flowrates are specified, as well as a list of all pipes and conveyors and their contents, material properties such as density, viscosity, particle size distribution, flow rates, pressures, temperatures, and materials of construction for the piping and unit operations. The process flow diagram is then used to develop a Piping and instrumentation diagram (P&ID) which includes pipe and conveyor sizing information to address the desired flowrates, process controls (such as tank level indications, material flow meters, weighing devices, motor speed controls, temperature and pressure indicators/ controllers, etc.). The P&ID is then used as a basis of design for developing the "system operation guide" or "functional design specification" which outlines the operation of the process. From the "P&ID", a proposed layout (general arrangement) of the process can be shown from an overhead view (plot plan) and a side view (elevation), and other engineering disciplines are involved such as civil engineers for site work (earth moving), foundation design, concrete slab design work, structural steel to support the equipment, etc.). All previous work is directed toward defining the scope of the project, then developing a cost estimate to get the design installed, and a schedule to communicate the timing needs for engineering, procurement, fabrication, installation, commissioning, startup, and ongoing production of the process. Depending on the needed accuracy of the cost estimate and schedule that is required, several iterations of designs are generally provided to customers or stakeholders who feedback their requirements and the process engineer incorporates these additional instructions and wants (scope revisions) into the overall design and additional cost estimates and Schedules are developed for funding approval. Following funding approval, the project is executed via project management.

The application of systematic computer-based methods to process engineering is process systems engineering.

Significant Accomplishments

Several accomplishments have been made in Process Systems Engineering:

- Process design: synthesis of energy recovery networks, synthesis of distillation systems (azeotropic), synthesis of reactor networks, hierarchical decomposition flowsheets, superstructure optimization, design multiproduct batch plants. Design of the production reactors for the production of plutonium, design of nuclear submarines.

- Process control: model predictive control, controllability measures, robust control, nonlinear control, statistical process control, process monitoring, thermodynamics-based control

- Process operations: scheduling process networks, multiperiod planning and optimization, data reconciliation, real-time optimization, flexibility measures, fault diagnosis

- Supporting tools: sequential modular simulation, equation based process simulation, AI/expert systems, large-scale nonlinear programming (NLP), optimization of differential algebraic equations (DAEs), mixed-integer nonlinear programming (MINLP), global optimization

History of Process Systems Engineering

Process systems engineering (PSE) is a relatively young area in chemical engineering. The first time that this term was used was in a Special Volume of the AIChE Symposium Series in 1961. However, it was not until 1982 when the first international symposium on this topic took place in Kyoto, Japan, that the term PSE started to become widely accepted.

The first textbook in the area was "Strategy of Process Engineering" by Dale F. Rudd and Charles C. Watson, Wiley, 1968. The Computing and Systems Technology (CAST) Division, Area 10 of AIChE, was founded in 1977 and currently has about 1200 members. CAST has four sections: Process Design, Process Control, Process Operations, and Applied Mathematics.

The first journal devoted to PSE was "Computers and Chemical Engineering," which appeared in 1977. The Foundations of Computer-Aided Process Design (FOCAPD) conference in 1980 in Henniker was one of the first meetings in a series on that topic in the PSE area. It is now accompanied by the successful series on Control (CPC), Operations (FOCAPO), and the world-wide series entitled Process Systems Engineering. The CACHE Corporation (Computer Aids for Chemical Engineering), which organizes these conferences, was initially launched by academics in 1970, motivated by the introduction of process simulation in the chemical engineering curriculum.

Roger W.H. Sargent from Imperial College was one of the pioneers in the area. PSE is an active area of research in many other countries, particularly in the United Kingdom, Germany, Japan, Korea, and China.

Plant Layout Study

A plant layout study is an engineering study used to analyze different physical configurations for a manufacturing plant. It is also known as Facilities Planning and Layout.

Introduction

The ability to design and operate manufacturing facilities that can quickly and effectively adapt to changing technological and market requirements is becoming increasingly important to the success of any manufacturing organization. In the face of shorter product life cycles, higher product variety, increasingly unpredictable demand, and shorter delivery times, manufacturing facilities dedicated to a single product line cannot be cost effective any longer. Investment efficiency now requires that manufacturing facilities be able to shift quickly from one product line to another without major retooling, resource reconfiguration, or replacement of equipment.

Investment efficiency also requires that manufacturing facilities be able to simultaneously make several products so that smaller volume products can be combined in a single facility and that fluctuations in product mixes and volumes can be more easily accommodated. In short, manufacturing facilities must be able to exhibit high levels of flexibility and robustness despite significant changes in their operating requirements.

In industry sectors, it is important to manufacture the products which have good quality and meet customers' demand. This action could be conducted under existing resources such as employees, machines and other facilities. However, plant layout improvement, could be one of the tools to response to increasing industrial productivity. Plant layout design has become a fundamental basis of today's industrial plants which can influence parts of work efficiency. It is needed to appropriately plan and position employees, materials, machines, equipment, and other manufacturing supports and facilities to create the most effective plant layout.

Product Considerations

The intended products to be manufactured influence the choice of layout.

References

- Oliver, David W.; Timothy P. Kelliher, James G. Keegan, Jr. (1997). Engineering Complex Systems with Models and Objects. McGraw-Hill. pp. 85–94. ISBN 0-07-048188-1.

- Gianni, Daniele; D'Ambrogio, Andrea; Tolk, Andreas, eds. (4 December 2014). Modeling and Simulation-Based Systems Engineering Handbook (1st ed.). CRC Press. p. 513. ISBN 9781466571457.

- Dhillon, Balbir S. (2006) Maintainability, Maintenance, and Reliability for Engineers, CRC Press, 2006, ISBN 978-0-8493-7243-7, ISBN 978-0-8493-7243-8;

- Mobley, Keith R. & Higgins, Lindley R. & Wikoff, Darrin J. (2008) Maintenance Engineering Handbook, McGraw-Hill Professional, Seventh Edition, 2008, ISBN 0-07-154646-4, ISBN 978-0-07-154646-1;

- business competitive edge perspective. Cairo: International Business Information Management Association (IBIMA),2010. s. 1881-1886. ISBN 978-0-9821489-4-5

- Karniel, Arie; Reich, Yoram (2011). Managing the Dynamic of New Product Development Processes. A new Product Lifecycle Management Paradigm. Springer. p. 13. ISBN 978-0-85729-569-9. Retrieved 25 February 2012.

- Malakooti, Behnam (2013). Operations and Production Systems with Multiple Objectives. John Wiley & Sons. ISBN 978-1-118-58537-5.

- Twede, Diana; Selke, Susan (December 2014). Cartons, Crates & Corrugated Board, handbook of paper & wood packaging technology, Second Edition (2nd ed.). Lancaster PA: DEStech Publications. ISBN 978-1-60595-135-5.

- Based on p. 475 of European Agency for Safety and Health at Work (2000): Monitoring the state of occupational safety and health in the European Union – Pilot Study, Bilbao, Spain: European Agency for Safety and Health at Work, ISBN 92-95007-00-X,

- Health and Safety Executive (2009): A Guide to Safety and Health Regulation in Great Britain. 4th edition. ISBN 978-0-7176-6319-4,

- Chaturvedi, Pradeep (2006-01-01). Challenges of Occupational Safety and Health: Thrust : Safety in Transportation.Concept Publishing Company. ISBN 9788180692840.

Theory and Planning of Industrial Engineering

There are many prevalent theories related to industrial engineering. The most significant out of them is the theory of constraints which has been lucidly covered in this chapter. Another extremely crucial aspect of industrial engineering is planning. This chapter throws light on the planning process involved in industrial engineering, especially in terms of material requirements.

Material Requirements Planning

Material requirements planning (MRP) is a production planning, scheduling, and inventory control system used to manage manufacturing processes. Most MRP systems are software-based, while it is possible to conduct MRP by hand as well.

An MRP system is intended to simultaneously meet three objectives:

- Ensure materials are available for production and products are available for delivery to customers.

- Maintain the lowest possible material and product levels in store

- Plan manufacturing activities, delivery schedules and purchasing activities.

History

Prior to MRP, and before computers dominated industry, reorder point (ROP)/reorder-quantity (ROQ) type methods like EOQ (economic order quantity) had been used in manufacturing and inventory management.

In 1964, as a response to the Toyota Manufacturing Program, Joseph Orlicky developed material requirements planning (MRP). The first company to use MRP was Black & Decker in 1964, with Dick Alban as project leader. Orlicky's 1975 book *Material Requirements Planning* has the subtitle *The New Way of Life in Production and Inventory Management*. By 1975, MRP was implemented in 700 companies. This number had grown to about 8,000 by 1981.

In 1983, Oliver Wight developed MRP into manufacturing resource planning (MRP II). In the 1980s, Joe Orlicky's MRP evolved into Oliver Wight's manufacturing resource planning (MRP II) which brings master scheduling, rough-cut capacity planning, capacity requirements planning,

S&OP in 1983 and other concepts to classical MRP. By 1989, about one third of the software industry was MRP II software sold to American industry ($1.2 billion worth of software).

The Scope of MRP in Manufacturing

Dependent Demand Vs Independent Demand

Independent demand is demand originating outside the plant or production system, while dependent demand is demand for components. The bill of materials (BOM) specifies the relationship between the end product (independent demand) and the components (dependent demand). MRP take as input the information contained in the BOM.

Functions

The basic functions of an MRP system include: inventory control, bill of material processing, and elementary scheduling. MRP helps organizations to maintain low inventory levels. It is used to plan manufacturing, purchasing and delivering activities.

"Manufacturing organizations, whatever their products, face the same daily practical problem - that customers want products to be available in a shorter time than it takes to make them. This means that some level of planning is required."

Companies need to control the types and quantities of materials they purchase, plan which products are to be produced and in what quantities and ensure that they are able to meet current and future customer demand, all at the lowest possible cost. Making a bad decision in any of these areas will make the company lose money. A few examples are given below:

- If a company purchases insufficient quantities of an item used in manufacturing (or the wrong item) it may be unable to meet contract obligations to supply products on time.

- If a company purchases excessive quantities of an item, money is wasted - the excess quantity ties up cash while it remains as stock and may never even be used at all.

- Beginning production of an order at the wrong time can cause customer deadlines to be missed.

MRP is a tool to deal with these problems. It provides answers for several questions:

- *What* items are required?

- *How many* are required?

- *When* are they required?...

MRP can be applied both to items that are purchased from outside suppliers and to sub-assemblies, produced internally, that are components of more complex items.

Data

The data that must be considered include:

- The *end item* (or items) being created. This is sometimes called independent demand, or Level "0" on BOM (bill of materials).

- How much is required at a time.

- When the quantities are required to meet demand.

- Shelf life of stored materials.

- Inventory status records. Records of *net* materials *available* for use already in stock (on hand) and materials on order from suppliers.

- Bills of materials. Details of the materials, components and sub-assemblies required to make each product.

- Planning data. This includes all the restraints and directions to produce the end items. This includes such items as: routing, labor and machine standards, quality and testing standards, pull/work cell and push commands, lot sizing techniques (i.e. fixed lot size, lot-for-lot, economic order quantity), scrap percentages, and other inputs.

Outputs

There are two outputs and a variety of messages/reports:

- Output 1 is the "Recommended Production Schedule" which lays out a detailed schedule of the required minimum start and completion dates, with quantities, for each step of the Routing and Bill Of Material required to satisfy the demand from the master production schedule (MPS).

- Output 2 is the "Recommended Purchasing Schedule". This lays out both the dates that the purchased items should be received into the facility *and* the dates that the purchase orders or blanket order release should occur to match the production schedules.

Messages and reports:

- Purchase orders. An order to a supplier to provide materials.

- Reschedule notices. These *recommend* cancelling, increasing, delaying or speeding up existing orders.

Methods to Find Order Quantities

Well-known methods to find order quantities are:

- Dynamic lot-sizing

- Silver–Meal heuristic

- Least-Unit-Cost heuristic

Mathematical Formulation

MRP can be expressed as an optimal control problem:

Initial conditions:

$$x_i'(0) = x_{i0}' \quad i=1,\ldots,J$$

Dynamics:

$$x_i'(t+1) = x_i'(t) + z_i(t) - d_i'(t) - \sum_{j \in Suc(i)} z_j(t + L_{ij}' - 1)$$

$$t=0,\ldots,T\text{-}1, i=1,\ldots,J$$

Constraints:

$$x_i'(t) \geq 0 \quad t=1,\ldots,T, i=1,\ldots,J$$

$$z_i(t) \geq 0 \quad t=0,\ldots,T\text{-}1, i=1,\ldots,J$$

Objective:

$$\min \sum_i \sum_{t=0}^{T-1} \left[k_i(t)\delta(z_i(t)) + c_i(t)z_i(t) \right] + \sum_i \sum_{t=1}^{T} h_i'(t)x_i'(t)$$

Where x' is local inventory (the state), z the order size (the control), d is local demand, k represents fixed order costs, c variable order costs, h local inventory holding costs. $\delta()$ is the Heaviside function. Changing the dynamics of the problem leads to a multi-item analogue of the dynamic lot-size model.

Problems with MRP Systems

- Integrity of the data. If there are any errors in the inventory data, the bill of materials (commonly referred to as 'BOM') data, or the master production schedule, then the output data will also be incorrect ("GIGO": garbage in, garbage out). Data integrity is also affected by inaccurate cycle count adjustments, mistakes in receiving input and shipping output, scrap not reported, waste, damage, box count errors, supplier container count errors, production reporting errors, and system issues. Many of these type of errors can be minimized by implementing pull systems and using bar code scanning. Most vendors in this type of system recommend at least 99% data integrity for the system to give useful results.

- Systems require that the user specify how long it will take for a factory to make a product from its component parts (assuming they are all available). Additionally, the system design also assumes that this "lead time" in manufacturing will be the same each time the item is made, without regard to quantity being made, or other items being made simultaneously in the factory.

- A manufacturer may have factories in different cities or even countries. It is not good for an MRP system to say that we do not need to order some material, because we have plenty of it thousands of miles away. The overall ERP system needs to be able to organize inventory and needs by individual factory and inter-communicate the needs in order to enable each factory to redistribute components to serve the overall enterprise. This means that other systems in the enterprise need to work properly, both before implementing an MRP system and in the future. For example, systems like variety reduction and engineering, which makes sure that product comes out right first time (without defects), must be in place.

- Production may be in progress for some part, whose design gets changed, with customer orders in the system for both the old design, and the new one, concurrently. The overall ERP system needs to have a system of coding parts such that the MRP will correctly calculate needs and tracking for both versions. Parts must be booked into and out of stores more regularly than the MRP calculations take place. Note, these other systems can well be manual systems, but must interface to the MRP. For example, a 'walk around' stock intake done just prior to the MRP calculations can be a practical solution for a small inventory (especially if it is an "open store").

- The other major drawback of MRP is that it fails to account for capacity in its calculations. This means it will give results that are impossible to implement due to manpower, machine or supplier capacity constraints. However this is largely dealt with by MRP II. Generally, MRP II refers to a system with integrated financials. An MRP II system can include finite or infinite capacity planning. But, to be considered a true MRP II system must also include financials. In the MRP II (or MRP2) concept, fluctuations in forecast data are taken into account by including simulation of the master production schedule, thus creating a long-term control. A more general feature of MRP2 is its extension to purchasing, to marketing and to finance (integration of all the functions of the company), ERP has been the next step.

Solutions to Data Integrity Issues

- Bill of material – The best practice is to physically verify the bill of material either at the production site or by disassembling the product.

- Cycle count – The best practice is to determine why a cycle count that increases or decreases inventory has occurred. Find the root cause and correct the problem from occurring again.

- Scrap reporting – This can be the most difficult area to maintain with any integrity. Start with isolating the scrap by providing scrap bins at the production site and then record the scrap from the bins on a daily basis. One benefit of reviewing the scrap on site is that preventive action can be taken by the engineering group.

- Receiving errors – Manual systems of recording what has been received are error prone. The best practice is to implement the system of receiving by ASN from the supplier. The supplier sends an ASN (advanced shipping notification). When the components are received into the facility, the ASN is processed and then company labels are created for each line item. The labels are affixed to each container and then scanned into the MRP system. Extra labels reveal a shortage from the shipment and too few labels reveal an over shipment. Some companies pay for ASN by reducing the time in processing accounts payable.

- Shipping errors – The container labels are printed from the shipper. The labels are affixed to the containers in a staging area or when they are loaded on the transport.

- Production reporting – The best practice is to use bar code scanning to enter production into inventory. A product that is rejected should be moved to an MRB (material review board) location. Containers that require sorting need to be received in reverse.

- Replenishment – The best replenishment practice is replacement using bar code scanning, or via pull system. Depending upon the complexity of the product, planners can actually order materials using scanning with a min-max system.

Demand Driven MRP

In 2011, the third edition of "Orlicky's Planning" introduced a new type of MRP called "demand driven MRP" (DDMRP). The new edition of the book was written, not by Orlicky himself (he died in 1986) but by Carol Ptak and Chad Smith at the invitation of McGraw Hill to update Orlicky's work.

Demand driven MRP is a multi-echelon formal planning and execution technique with five distinct components:

1. Strategic inventory positioning – The first question of effective inventory management is not, "how much inventory should we have?" Nor is it, "when should we make or buy something?" The most fundamental question to ask in today's manufacturing environments is, "given our system and environment, where should we place inventory to have the best protection?" Inventory is like a break wall to protect boats in a marina from the roughness of incoming waves. Out on the open ocean the break walls have to be 50–100 feet tall, but in a small lake the break walls are only a couple feet tall. In a glassy smooth pond no break wall is necessary.

2. Buffer profiles and level – Once the strategically replenished positions are determined, the actual levels of those buffers have to be initially set. Based on several factors, different materials and parts behave differently (but many also behave nearly the same). DDMRP calls for the grouping of parts and materials chosen for strategic replenishment and that behave similarly into "buffer profiles." Buffer profiles take into account important factors including lead time (relative to the environment), variability (demand or supply), whether the part is made or bought or distributed and whether there are significant order multiples involved. These buffer profiles are made up of "zones" that produce a unique buffer picture for each part as their respective individual part traits are applied to the group traits.

3. Dynamic adjustments – Over the course of time, group and individual traits can and will change as new suppliers and materials are used, new markets are opened and/or old markets deteriorate and manufacturing capacities and methods change. Dynamic buffer levels allow the company to adapt buffers to group and individual part trait changes over time through the use of several types of adjustments. Thus, as more or less variability is encountered or as a company's strategy changes these buffers adapt and change to fit the environment.

4. Demand-driven planning – takes advantage of the sheer computational power of today's hardware and software. It also takes advantage of the new demand-driven or pull-based approaches. When these two elements are combined then there is the best of both worlds;

relevant approaches and tools for the way the world works today *and* a system of routine that promotes better and quicker decisions and actions at the planning and execution level.

5. Highly visible and collaborative execution – Simply launching purchase orders (POs), manufacturing orders (MOs) and transfer orders (TOs) from any planning system does not end the materials and order management challenge. These POs, MOs and TOs have to be effectively managed to synchronize with the changes that often occur within the "execution horizon." The execution horizon is the time from which a PO, MO or TO is opened until the time it is closed in the system of record. DDMRP defines a modern, integrated and greatly needed system of execution for all part categories in order to speed the proliferation of relevant information and priorities throughout an organization and supply chain.

These five components work together to greatly dampen, if not eliminate, the nervousness of traditional MRP systems and the bullwhip effect in complex and challenging environments. Many claim have been made by the consultancy company that is marketing DDMRP, including the following: In utilizing these approaches, planners will no longer have to try to respond to every single message for every single part that is off by even one day. This approach provides real information about those parts that are truly at risk of negatively impacting the planned availability of inventory. DDMRP sorts the significant few items that require attention from the many parts that are being managed. Under the DDMRP approach, fewer planners can make better decisions more quickly. That means companies will be better able to leverage their working and human capital as well as the huge investments they have made in information technology. One down-side, however, is that DDMRP can not run on the majority of MRPII/ERP systems in use today, so companies that wish to use it have to replace their current system with a 'certified' system, only one of which currently exists.

DDMRP has been successfully applied to a variety of environments including CTO (configure to order), MTS (make to stock), MTO (make to order) and ETO (engineer to order). (Most DDMRP consultants, however, believe that DDMRP only produces better results that standard MRP in MTF (make to forecast) environments.) The methodology is applied differently in each environments but the five step process remains the same. DDMRP leverages knowledge from theory of constraints (TOC), traditional MRP & DRP, Six Sigma and lean. It is effectively an amalgam of MRP and kanban techniques. As such, it shares the strengths of both but also the weaknesses of both and, in consequence, it has not been widely implemented..

Theory of Constraints

The theory of constraints (TOC) is a management paradigm that views any manageable system as being limited in achieving more of its goals by a very small number of constraints. There is always at least one constraint, and TOC uses a focusing process to identify the constraint and restructure the rest of the organization around it. TOC adopts the common idiom "a chain is no stronger than its weakest link." This means that processes, organizations, etc., are vulnerable because the weakest person or part can always damage or break them or at least adversely affect the outcome.

History

The theory of constraints (TOC) is an overall management philosophy introduced by Eliyahu M. Goldratt in his 1984 book titled *The Goal*, that is geared to help organizations continually achieve their goals. Goldratt adapted the concept to project management with his book *Critical Chain*, published in 1997.

An earlier propagator of a similar concept was Wolfgang Mewes in Germany with publications on *power-oriented management theory* (Machtorientierte Führungstheorie, 1963) and following with his *Energo-Kybernetic System (EKS, 1971)*, later renamed *Engpasskonzentrierte Strategie* as a more advanced *theory of bottlenecks*. The publications of Wolfgang Mewes are marketed through the FAZ Verlag, publishing house of the German newspaper *Frankfurter Allgemeine Zeitung*. However, the paradigm *Theory of constraints* was first used by Goldratt.

Key Assumption

The underlying premise of the theory of constraints is that organizations can be measured and controlled by variations on three measures: throughput, operational expense, and inventory. Inventory is all the money that the system has invested in purchasing things which it intends to sell. Operational expense is all the money the system spends in order to turn inventory into throughput. Throughput is the rate at which the system generates money through sales.

Before the goal itself can be reached, necessary conditions must first be met. These typically include safety, quality, legal obligations, etc. For most businesses, the goal itself is to make money. However, for many organizations and non-profit businesses, making money is a necessary condition for pursuing the goal. Whether it is the goal or a necessary condition, understanding how to make sound financial decisions based on throughput, inventory, and operating expense is a critical requirement.

The Five Focusing Steps

Theory of constraints is based on the premise that the rate of goal achievement by a goal-oriented system (i.e., the system's throughput) is limited by at least one constraint.

The argument by reductio ad absurdum is as follows: If there was nothing preventing a system from achieving higher throughput (i.e., more goal units in a unit of time), its throughput would be infinite — which is impossible in a real-life system.

Only by increasing flow through the constraint can overall throughput be increased.

Assuming the goal of a system has been articulated and its measurements defined, the steps are:

1. Identify the system's constraint(s).

2. Decide how to exploit the system's constraint(s).

3. Subordinate everything else to the above decision(s).

4. Elevate the system's constraint(s).

5. Warning! If in the previous steps a constraint has been broken, go back to step 1, but do not allow inertia to cause a system's constraint.

The goal of a commercial organization is: "Make more money now and in the future", and its measurements are given by throughput accounting as: throughput, inventory, and operating expenses.

The five focusing steps aim to ensure ongoing improvement efforts are centered on the organization's constraint(s). In the TOC literature, this is referred to as the *process of ongoing improvement* (POOGI).

These focusing steps are the key steps to developing the specific applications mentioned below.

Constraints

A constraint is anything that prevents the system from achieving its goal. There are many ways that constraints can show up, but a core principle within TOC is that there are not tens or hundreds of constraints. There is at least one but at most only a few in any given system. Constraints can be internal or external to the system. An internal constraint is in evidence when the market demands more from the system than it can deliver. If this is the case, then the focus of the organization should be on discovering that constraint and following the five focusing steps to open it up (and potentially remove it). An external constraint exists when the system can produce more than the market will bear. If this is the case, then the organization should focus on mechanisms to create more demand for its products or services.

Types of (internal) constraints

* Equipment: The way equipment is currently used limits the ability of the system to produce more salable goods/services.

* People: Lack of skilled people limits the system. Mental models held by people can cause behaviour that becomes a constraint.

* Policy: A written or unwritten policy prevents the system from making more.

The concept of the constraint in Theory of Constraints is analogous to but differs from the constraint that shows up in mathematical optimization. In TOC, the constraint is used as a focusing mechanism for management of the system. In optimization, the constraint is written into the mathematical expressions to limit the scope of the solution (X can be no greater than 5).

Please note: organizations have many problems with equipment, people, policies, etc. (A breakdown is just that – a breakdown – and is not a constraint in the true sense of the TOC concept) The constraint is the limiting factor that is preventing the organization from getting more throughput (typically, revenue through sales) even when nothing goes wrong.

Breaking a Constraint

If a constraint's throughput capacity is elevated to the point where it is no longer the system's limiting factor, this is said to "break" the constraint. The limiting factor is now some other part of the system, or may be external to the system (an external constraint).

Buffers

Buffers are used throughout the theory of constraints. They often result as part of the exploit and subordinate steps of the five focusing steps. Buffers are placed before the governing constraint, thus ensuring that the constraint is never starved. Buffers are also placed behind the constraint to prevent downstream failure from blocking the constraint's output. Buffers used in this way protect the constraint from variations in the rest of the system and should allow for normal variation of processing time and the occasional upset (Murphy) before and behind the constraint.

Buffers can be a bank of physical objects before a work center, waiting to be processed by that work center. Buffers ultimately buy you time, as in the time before work reaches the constraint and are often verbalized as time buffers. There should always be enough (but not excessive) work in the time queue before the constraint and adequate offloading space behind the constraint.

Buffers are *not* the small queue of work that sits before every work center in a Kanban system although it is similar if you regard the assembly line as the governing constraint. A prerequisite in the theory is that with one constraint in the system, all other parts of the system must have sufficient capacity to keep up with the work at the constraint and to catch up if time was lost. In a balanced line, as espoused by Kanban, when one work center goes down for a period longer than the buffer allows, then the entire system must wait until that work center is restored. In a TOC system, the only situation where work is in danger is if the constraint is unable to process (either due to malfunction, sickness or a "hole" in the buffer – if something goes wrong that the time buffer can not protect).

Buffer management, therefore, represents a crucial attribute of the theory of constraints. There are many ways to apply buffers, but the most often used is a visual system of designating the buffer in three colors: green (okay), yellow (caution) and red (action required). Creating this kind of visibility enables the system as a whole to align and thus subordinate to the need of the constraint in a holistic manner. This can also be done daily in a central operations room that is accessible to everybody.

Plant Types

There are four primary types of plants in the TOC lexicon. Draw the flow of material from the bottom of a page to the top, and you get the four types. They specify the general flow of materials through a system, and they provide some hints about where to look for typical problems. The four types can be combined in many ways in larger facilities.

- I-plant: Material flows in a sequence, such as in an assembly line. The primary work is done in a straight sequence of events (one-to-one). The constraint is the slowest operation.

- A-plant: The general flow of material is many-to-one, such as in a plant where many sub-assemblies converge for a final assembly. The primary problem in A-plants is in synchronizing the converging lines so that each supplies the final assembly point at the right time.

- V-plant: The general flow of material is one-to-many, such as a plant that takes one raw material and can make many final products. Classic examples are meat rendering plants or a steel manufacturer. The primary problem in V-plants is "robbing" where one operation (A) immediately after a diverging point "steals" materials meant for the other operation (B). Once the material has been processed by A, it cannot come back and be run through B without significant rework.

- T-plant: The general flow is that of an I-plant (or has multiple lines), which then splits into many assemblies (many-to-many). Most manufactured parts are used in multiple assemblies and nearly all assemblies use multiple parts. Customized devices, such as computers, are good examples. T-plants suffer from both synchronization problems of A-plants (parts aren't all available for an assembly) and the robbing problems of V-plants (one assembly steals parts that could have been used in another).

For non-material systems, one can draw the flow of work or the flow of processes and arrive at similar basic structures. A project, for example, is an A-shaped sequence of work, culminating in a delivered project.

Applications

The focusing steps, this process of ongoing improvement, have been applied to manufacturing, project management, supply chain/distribution generated specific solutions. Other tools (mainly the "thinking process") also led to TOC applications in the fields of marketing and sales, and finance. The solution as applied to each of these areas are listed below.

Operations

Within manufacturing operations and operations management, the solution seeks to pull materials through the system, rather than push them into the system. The primary methodology used is drum-buffer-rope (DBR) and a variation called simplified drum-buffer-rope (S-DBR).

Drum-buffer-rope is a manufacturing execution methodology based on the fact the output of a system can only be the same as the output at the constraint of the system. Any attempt to produce more than what the constraint can process just leads to excess inventory piling up. The method is named for its three components. The *drum* is the physical constraint of the plant: the work center or machine or operation that limits the ability of the entire system to produce more. The rest of the plant follows the beat of the drum. Schedule at the drum decides what the system should produce, in what sequence to produce and how much to produce. They make sure the drum has work and that anything the drum has processed does not get wasted.

The *buffer* protects the drum, so that it always has work flowing to it. Buffers in DBR provide the additional lead time beyond the required set up and process times, for materials in the product flow. Since these buffers have time as their unit of measure, rather than quantity of material, this makes the priority system operate strictly based on the time an order is expected to be at the drum. Each work order will have a remaining buffer status that can be calculated. Based on this buffer status work orders can be color coded into Red, Yellow and Green. The red orders have the highest priority and must worked on first since they have penetrated most into their buffers followed by yellow and green. As time evolves this buffer status might change and the color assigned to the particular work order change with it.

Traditional DBR usually calls for buffers at several points in the system: the constraint, synchronization points and at shipping. S-DBR has a buffer at shipping and manages the flow of work across the drum through a load planning mechanism.

The *rope* is the work release mechanism for the plant. Orders are released to the shop floor at one "buffer time" before they are due. In other words, if the buffer is 5 days, the order is released 5 days before it is due at the constraint. Putting work into the system earlier than this buffer time is likely to generate too-high work-in-process and slow down the entire system.

High-Speed Automated Production Lines

Automated production lines that are used in the beverage industry to fill containers usually have several machines executing parts of the complete process, from filling primary containers to secondary packaging and palletisation. These machines operate at different speeds and capacities and have varying efficiency levels.

To be able to maximize the throughput, the production line usually has a designed constraint. This constraint is typically the slowest and usually the most expensive machine on the line. The overall throughput of the line is determined by this machine. All other machines can operate faster and are connected by conveyors.

The conveyors usually have the ability to buffer product. In the event of a stoppage at a machine other than the constraint, the conveyor can buffer the product enabling the constraint machine to keep on running.

A typical line setup is such that in normal operation the upstream conveyors from the constraint machine are always run full to prevent starvation at the constraint and the downstream conveyors are run empty to prevent a back up at the constraint. The overall aim is to prevent minor stoppages at the machines from impacting the constraint.

For this reason as the machines get further from the constraint, they have the ability to run faster than the previous machine and this creates a V curve.

Supply Chain and Logistics

In general, the solution for supply chains is to create flow of inventory so as to ensure greater availability and to eliminate surpluses.

The TOC distribution solution is effective when used to address a single link in the supply chain and more so across the entire system, even if that system comprises many different companies. The purpose of the TOC distribution solution is to establish a decisive competitive edge based on extraordinary availability by dramatically reducing the damages caused when the flow of goods is interrupted by shortages and surpluses.

This approach uses several new rules to protect availability with less inventory than is conventionally required. Before explaining these new rules, the term Replenishment Time must be defined. Replenishment Time (RT) is the sum of the delay, after the first consumption following a delivery, before an order is placed plus the delay after the order is placed until the ordered goods arrive at the ordering location.

1. Inventory is held at an aggregation point(s) as close as possible to the source. This approach ensures smoothed demand at the aggregation point, requiring proportionally less inventory. The distribution centers holding the aggregated stock are able to ship goods downstream to the next link in the supply chain much more quickly than a make-to-order manufacturer can.

2. Following this rule may result in a make-to-order manufacturer converting to make-to-stock. The inventory added at the aggregation point is significantly less than the inventory reduction downstream.

3. In all stocking locations, initial inventory buffers are set which effectively create an upper limit of the inventory at that location. The buffer size is equal to the maximum expected consumption within the average RT, plus additional stock to protect in case a delivery is late. In other words, there is no advantage in holding more inventory in a location than the amount that might be consumed before more could be ordered and received. Typically, the sum of the on hand value of such buffers are 25–75% less than currently observed average inventory levels.

4. Once buffers have been established, no replenishment orders are placed as long as the quantity inbound (already ordered but not yet received) plus the quantity on hand are equal to or greater than the buffer size. Following this rule causes surplus inventory to be bled off as it is consumed.

5. For any reason, when on hand plus inbound inventory is less than the buffer, orders are placed as soon as practical to increase the inbound inventory so that the relationship On Hand + Inbound = Buffer is maintained.

6. To ensure buffers remain correctly sized even with changes in the rates of demand and replenishment, a simple recursive algorithm called Buffer Management is used. When the on hand inventory level is in the upper third of the buffer for a full RT, the buffer is reduced by one third (and don't forget rule 3). Alternatively, when the on hand inventory is in the bottom one third of the buffer for too long, the buffer is increased by one third (and don't forget rule 4). The definition of "too long" may be changed depending on required service levels, however, a rule of thumb is 20% of the RT. Moving buffers up more readily than down is supported by the usually greater damage caused by shortages as compared to the damage caused by surpluses.

Once inventory is managed as described above, continuous efforts should be undertaken to reduce RT, late deliveries, supplier minimum order quantities (both per SKU and per order) and customer order batching. Any improvements in these areas will automatically improve both availability and inventory turns, thanks to the adaptive nature of Buffer Management.

A stocking location that manages inventory according to the TOC should help a non-TOC customer (downstream link in a supply chain, whether internal or external) manage their inventory according to the TOC process. This type of help can take the form of a vendor managed inventory (VMI). The TOC distribution link simply extends its buffer sizing and management techniques to its customers' inventories. Doing so has the effect of smoothing the demand from the customer and reducing order sizes per SKU. VMI results in better availability and inventory turns for both supplier and customer. More than that, the benefits to the non-TOC customers are sufficient to meet the purpose of capitalizing on the decisive competitive edge by giving the customer a powerful reason to be more loyal and give more business to the upstream link. When the end consumers buy more the whole supply chain sells more.

One caveat should be considered. Initially and only temporarily, the supply chain or a specific link may sell less as the surplus inventory in the system is sold. However, the immediate sales lift due to improved availability is a countervailing factor. The current levels of surpluses and shortages make each case different.

Finance and Accounting

The solution for finance and accounting is to apply holistic thinking to the finance application. This has been termed throughput accounting. Throughput accounting suggests that one examine the impact of investments and operational changes in terms of the impact on the throughput of the business. It is an alternative to cost accounting.

The primary measures for a TOC view of finance and accounting are: throughput, operating expense and investment. Throughput is calculated from sales minus "totally variable cost", where totally variable cost is usually calculated as the cost of raw materials that go into creating the item sold.

Project Management

Critical Chain Project Management (CCPM) are utilized in this area. CCPM is based on the idea that all projects look like A-plants: all activities converge to a final deliverable. As such, to protect the project, there must be internal buffers to protect synchronization points and a final project buffer to protect the overall project.

Marketing and Sales

While originally focused on manufacturing and logistics, TOC has expanded lately into sales management and marketing. Its role is explicitly acknowledged in the field of sales process engineering. For effective sales management one can apply Drum Buffer Rope to the sales process similar to the way it is applied to operations. This technique is appropriate when your constraint is in the sales process itself or you just want an effective sales management technique and includes the topics of funnel management and conversion rates.

Thinking Processes

The thinking processes are a set of tools to help managers walk through the steps of initiating and implementing a project. When used in a logical flow, the thinking processes help walk through a buy-in process:

1. Gain agreement on the problem

2. Gain agreement on the direction for a solution

3. Gain agreement that the solution solves the problem

4. Agree to overcome any potential negative ramifications

5. Agree to overcome any obstacles to implementation

TOC practitioners sometimes refer to these in the negative as working through *layers of resistance* to a change.

Recently, the *current reality tree* (CRT) and *future reality tree* (FRT) have been applied to an argumentative academic paper.

Despite its origins as a manufacturing approach (Goldratt & Cox, The Goal: A process of Ongoing Improvement, 1992), Goldratt's Theory of Constraints (TOC) methodology is now regarded as a systems methodology with strong foundations in the hard sciences (Mabin, 1999). Through its tools for convergent thinking and synthesis, the "Thinking processes", which underpin the entire TOC methodology, help identify and manage constraints and guide continuous improvement and change in organizations (Dettmer H. , 1998).

The process of change requires the identification and acceptance of core issues; the goal and the means to the goal. This comprehensive set of logical tools can be used for exploration, solution development and solution implementation for individuals, groups or organizations. Each tool has a purpose and nearly all tools can be used independently (Cox & Spencer, 1998). Since these thinking tools are designed to address successive "layers of resistance" and enable communication, it expedites securing "buy in" of groups. While CRT (current reality tree) represents the undesirable effects of the current situation, the FRT (the future reality tree), NBR (negative branch) help people plan and understand the possible results of their actions. The PRT (Perquisite tree) and TRT (transition tree) are designed to build collective buy in and aid in the Implementation phase. The logical constructs of these tools or diagrams are the necessary condition logic, the sufficient cause logic and the strict logic rules that are used to validate cause-effect relationships which are modelled with these tools (Dettmer W. , 2006).

A summary of these tools, the questions they help answer and the associated logical constructs used is presented in the table below.

	Sufficient thinking "If……. then"	Necessary Thinking "In order to...we must"
What to change?	Current Reality Tree	
What to change to?	Future Reality Tree Negative Branch Reservations	Evaporating cloud
How to change?	Transition Tree	Perquisite Tree

TOC Thinking Process Tools: Use of these tools are based on the fundamental beliefs of TOC that organizations a) are inherently simple (interdependencies exist in organizations) b) desire inherent harmony (win – win solutions are possible) c) are inherently good (people are good) and have inherent potential (people and organizations have potential to do better) (Goldratt E. , 2009). In the book "Through the clouds to solutions" Jelena Fedurko (Fedurko, 2013) states that the major areas for application of TP tools as:

- To create and enhance thinking and learning skills
- To make better decisions
- To develop responsibility for one's own actions through understanding their consequences
- To handle conflicts with more confidence and win-win outcomes
- To correct behavior with undesirable consequences
- Assist in evaluating conditions for achieving a desired outcome
- To assist in peer mediation
- To assist in relationship between subordinates and bosses

Development and Practice

TOC was initiated by Goldratt, who until his recent death was still the main driving force behind the development and practice of TOC. There is a network of individuals and small companies loosely coupled as practitioners around the world. TOC is sometimes referred to as "constraint management". TOC is a large body of knowledge with a strong guiding philosophy of growth.

Criticism

Criticisms that have been leveled against TOC include:

Claimed Suboptimality of Drum-Buffer-Rope

While TOC has been compared favorably to linear programming techniques, D. Trietsch from University of Auckland argues that DBR methodology is inferior to competing methodologies. Linhares, from the Getulio Vargas Foundation, has shown that the TOC approach to establishing an optimal product mix is unlikely to yield optimum results, as it would imply that P=NP.

Unacknowledged Debt

Duncan (as cited by Steyn) says that TOC borrows heavily from systems dynamics developed by Forrester in the 1950s and from statistical process control which dates back to World War II. And Noreen Smith and Mackey, in their independent report on TOC, point out that several key concepts in TOC "have been topics in management accounting textbooks for decades."

People claim Goldratt's books fail to acknowledge that TOC borrows from more than 40 years of previous management science research and practice, particularly from program evaluation and review technique/critical path method (PERT/CPM) and the just in time strategy. A rebuttal to these criticisms is offered in Goldratt's "What is the *Theory of Constraints* and How Should it be Implemented?", and in his audio program, "Beyond The Goal". In these, Goldratt discusses the history of disciplinary sciences, compares the strengths and weaknesses of the various disciplines, and acknowledges the sources of information and inspiration for the thinking processes and critical chain methodologies. Articles published in the now-defunct Journal of *Theory of Constraints* referenced foundational materials. Goldratt published an article and gave talks with the title "Standing on the Shoulders of Giants" in which he gives credit for many of the core ideas of Theory of Constraints. Goldratt has sought many times to show the correlation between various improvement methods. However, many Goldratt adherents often denigrate other methodologies as inferior to TOC.

Goldratt has been criticized on lack of openness in his theories, an example being him not releasing the algorithm he used for the Optimum Performance Training system. Some view him as unscientific with many of his theories, tools and techniques not being a part of the public domain, rather a part of his own framework of profiting on his ideas. According to Gupta and Snyder (2009), despite being recognized as a genuine management philosophy nowadays, TOC has yet failed to demonstrate its effectiveness in the academic literature and as such cannot be considered academically worthy enough to be called a widely recognized theory. TOC needs more case studies that prove a connection between implementation and improved financial performance. Nave (2002) argues that TOC does not take employees into account and fails to empower them in the production process. He also states that TOC fails to address unsuccessful policies as constraints. In contrast, Mukherjee and Chatterjee (2007) state that much of the criticism of Goldratt's work has been focused on the lack of rigour in his work, but not of the bottleneck approach, which are two different aspects of the issue.

Certification and Education

The Theory of Constraints International Certification Organization (TOCICO) is an independent not-for-profit incorporated society that sets exams to ensure a consistent standard of competence. It is overseen by a board of academic and industrial experts. It also hosts an annual international conference. The work presented at these conferences constitutes a core repository of the current knowledge.

References

- Malakooti, Behnam (2013). Operations and Production Systems with Multiple Objectives. John Wiley & Sons. ISBN 978-1-118-58537-5.
- Waldner, Jean-Baptiste (1992). "CIM: Principles of Computer Integrated Manufacturing". Chichester: John Wiley & Sons: 46. ISBN 0-471-93450-X.
- Cox, Jeff; Goldratt, Eliyahu M. (1986). The goal: a process of ongoing improvement. [Croton-on-Hudson, NY]: North River Press. ISBN 0-88427-061-0.
- Goldratt, Eliyahu; Fox, Robert (1986). The Race. [Croton-on-Hudson, NY]: North River Press. p. 179. ISBN 978-0-88427-062-1.
- Eric Noreen; Debra Smith; James T. Mackey (1995). The Theory of Constraints and its implications for Management Accounting. North River Press. ISBN 0-88427-116-1.
- Paul H. Selden (1997). Sales Process Engineering: A Personal Workshop. Milwaukee, WI: ASQ Quality Press. pp. 33–35, 264–268. ISBN 0-87389-418-9.
- Goldratt, Eliyahu M. (2009). "Standing on the shoulders of giants: production concepts versus production applications. The Hitachi Tool Engineering example.". Gestão & produção. 16 (3): 333–343. Retrieved 2015-12-19.

Methods and Processes of Industrial Engineering

In order to understand any branch of engineering it is fundamental to understand its varied methods and processes. This chapter will provide an in-depth knowledge about the methods and processes involved in industrial engineering which result in higher efficiency and output. Topics like methods engineering, flow process chart, work measurement, work sampling, etc. are thoroughly discussed in this chapter.

Methods Engineering

Methods engineering is a subspecialty of industrial engineering and manufacturing engineering concerned with human integration in industrial production processes.

Overview

Alternatively it can be described as the design of the productive process in which a person is involved. The task of the Methods engineer is to decide where humans will be utilized in the process of converting raw materials to finished products and how workers can most effectively perform their assigned tasks. The terms operation analysis, work design and simplification, and methods engineering and corporate re-engineering are frequently used interchangeably.

Lowering costs and increasing reliability and productivity are the objectives of methods engineering. These objectives are met in a five step sequence as follows: Project selection, data acquisition and presentation, data analysis, development of an ideal method based on the data analysis and, finally, presentation and implementation of the method.

Methods Engineering Topics

Project Selection

Methods engineers typically work on projects involving new product design, products with a high cost of production to profit ratio, and products associated with having poor quality issues. Different methods of project selection include the Pareto analysis, fish diagrams, Gantt charts, PERT charts, and job/work site analysis guides.

Data Acquisition and Presentation

Data that needs to be collected are specification sheets for the product, design drawings, quantity and delivery requirements, and projections as to how the product will perform or has performed in the market. The Gantt process chart can assist in the analysis of the man to machine interaction and it can aid in establishing the optimum number of workers and machines subject to the financial constraints of the operation. A flow diagram is frequently employed to represent the manufacturing process associated with the product.

Data Analysis

Data analysis enables the methods engineer to make decisions about several things, including: purpose of the operation, part design characteristics, specifications and tolerances of parts, materials, manufacturing process design, setup and tooling, working conditions, material handling, plant layout, and workplace design. Knowing the specifics (who, what, when, where, why, and how) of product manufacturing assists in the development of an optimum manufacturing method.

Ideal Method Development

Equations of synchronous and random servicing as well as line balancing are used to determine the ideal worker to machine ratio for the process or product chosen. Synchronous servicing is defined as the process where a machine is assigned to more than one operator, and the assigned operators and machine are occupied during the whole operating cycle. Random servicing of a facility, as the name indicates, is defined as a servicing process with a random time of occurrence and need of servicing variables. Line balancing equations determine the ideal number of workers needed on a production line to enable it to work at capacity.

Presentation and Methods Implementation

The industrial process or operation can be optimized using a variety of available methods. Each method design has its advantages and disadvantages. The best overall method is chosen using selection criteria and concepts involving value engineering, cost-benefit analysis, crossover charts, and economic analysis. The outcome of the selection process is then presented to the company for implementation at the plant. This last step involves "selling the idea" to the company brass, a skill the methods engineer must develop in addition to the normal engineering qualifications.

Flow Process Chart

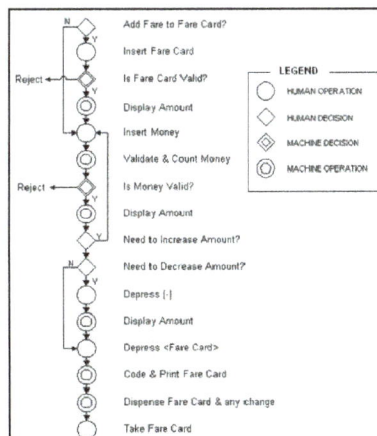

Subway Fare Card Machine Flow Process Chart (which doesn't use the ASME standard set of symbols).

The flow process chart in industrial engineering is a graphical and symbolic representation of the processing activities performed on the work piece.

History

The first structured method for documenting process flow, e.g. in flow shop scheduling, the flow process chart, was introduced by Frank and Lillian Gilbreth to members of ASME in 1921 as the presentation "Process Charts, First Steps in Finding the One Best Way to Do Work". The Gilbreths' tools quickly found their way into industrial engineering curricula.

In the early 1930s, an industrial engineer, Allan H. Mogensen began training business people in the use of some of the tools of industrial engineering at his Work Simplification Conferences in Lake Placid, New York. A 1944 graduate of Mogensen's class, Art Spinanger, took the tools back to Procter and Gamble where he developed their Deliberate Methods Change Program. Another 1944 graduate, Ben S. Graham, Director of Formcraft Engineering at Standard Register Corporation, adapted the flow process chart to information processing with his development of the multi-flow process chart to display multiple documents and their relationships.

Symbols

In 1947, ASME adopted the following symbol set derived from Gilbreth's original work as the ASME Standard for Process Charts.

Symbol	Letter	Description
O	O	Operation
□	I	Inspection
→	M	Move
D	D	Delay
∇	S	Storage

- Operation: to change the physical or chemical characteristics of the material.
- Inspection: to check the quality or the quantity of the material.
- Move: transporting the material from one place to another.
- Delay: when material cannot go to the next activity.
- Storage: when the material is kept in a safe location.

When To Use It

- It is used when observing a physical process, to record actions as they happen and thus get an accurate description of the process.
- It is used when analyzing the steps in a process, to help identify and eliminate waste - thus, it is a phenomenal tool when it comes to efficiency planning.
- It is used when the process is mostly sequential, containing few decisions.

Work Measurement

Work measurement is the application of techniques designed to establish the time for an average worker to carry out a specified manufacturing task at a defined level of performance. It is concerned with the length of time it takes to complete a work task assigned to a specific job .

Usage

Work measurement helps to uncover non-standardization that exist in the workplace and non-value adding activities and waste. A work has to be measured for the following reasons:

1. To discover and eliminate lost or ineffective time.

2. To establish standard times for performance measurement.

3. To measure performance against realistic expectations.

4. To set operating goals and objectives.

Techniques of Work Measurement

- time study

- Predetermined motion time systems

- Standard Data

- Work sampling

The Purpose of Work Measurement

Method study is the principal technique for reducing the work involved, primarily by eliminating unnecessary movement on the part of material or operatives and by substituting good methods for poor ones. Work measurement is concerned with investigating, reducing and subsequently eliminating ineffective time, that is time during which no effective work is being performed, whatever the cause.

Work measurement, as the name suggests, provides management with a means of measuring the time taken in the performance of an operation or series of operations in such a way that ineffective time is shown up and can be separated from effective time. In this way its existence, nature and extent become known where previously they were concealed within the total.

The Uses of Work Measurement

Revealing existing causes of ineffective time through study, important though it is, is perhaps less important in the long term than the setting of sound time standards, since these will continue to apply as long as the work to which they refer continues to be done. They will also show up any ineffective time or additional work which may occur once they have been established.

In the process of setting standards it may be necessary to use work measurement:

To compare the efficiency of alternative methods. Other conditions being equal, the method which takes the least time will be the best method. To balance the work of members of teams, in association with multiple activity charts, so that, as nearly as possible, each member has a task taking an equal time to perform. To determine, in association with man and machine multiple activity charts, the number of machines an operative can run.

The time standards, once set, may then be used:

To provide information on which the planning and scheduling of production can be based, including the plant and labour requirements for carrying out the programme of work and the utilisation of available capacity. To provide information on which estimates for tenders, selling prices and delivery promises can be based. To set standards of machine utilisation and labour performance which can be used for any of the above purposes and as a basis for incentive schemes. To provide information for labour-cost control and to enable standard costs to be fixed and maintained. It is thus clear that work measurement provides the basic information necessary for all the activities of organising and controlling the work of an enterprise in which the time element plays a part. Its uses in connection with these activities will be more clearly seen when we have shown how the standard time is obtained.

Techniques of Work Measurement

The following are the principal techniques by which work measurement is carried out:

1. Time study

2. Activity sampling

3. Predetermined motion time systems

4. Synthesis from standard data

5. Estimating

6. Analytical estimating

7. Comparative estimating

Of these techniques we shall concern ourselves primarily with time study, since it is the basic technique of work measurement. Some of the other techniques either derive from it or are variants of it.

1. Time Study

Time Study consists of recording times and rates of work for elements of a specified job carried out under specified conditions to obtain the time necessary to carry out a job at a defined level of performance.

In this technique the job to be studied is timed with a stopwatch, rated, and the Basic Time calculated.

1.1 Requirements for Effective Time Study

The requirements for effective time study are:

a. Co-operation and goodwill b. Defined job c. Defined method d. Correct normal equipment e. Quality standard and checks f. Experienced qualified motivated worker g. Method of timing h. Method of assessing relative performance i. Elemental breakdown j. Definition of break points k. Recording media

One of the most critical requirements for time study is that of elemental breakdown. There are some general rules concerning the way in which a job should be broken down into elements. They include the following. Elements should be easily identifiable, with definite beginnings and endings so that, once established, they can be repeatedly recognised. These points are known as the break points and should be clearly described on the study sheet. Elements should be as short as can be conveniently timed by the observer. As far as possible, elements - particularly manual ones - should be chosen so that they represent naturally unified and distinct segments of the operation.

1.2 Performance Rating

Time Study is based on a record of observed times for doing a job together with an assessment by the observer of the speed and effectiveness of the worker in relation to the observer's concept of Standard Rating.

This assessment is known as rating, the definition being given in BS 3138 (1979):

The numerical value or symbol used to denote a rate of working.

Standard rating is also defined (in this British Standard BS3138) as:

"The rating corresponding to the average rate at which qualified workers will naturally work, provided that they adhere to the specified method and that they are motivated to apply themselves to their work. If the standard rating is consistently maintained and the appropriate relaxation is taken, a qualified worker will achieve standard performance over the working day or shift."

Industrial engineers use a variety of rating scales, and one which has achieved wide use is the British Standards Rating Scale which is a scale where 0 corresponds to no activity and 100 corresponds to standard rating. Rating should be expressed as 'X' BS.

Below is an illustration of the Standard Scale:

Rating Walking Pace

0 no activity 50 very slow 75 steady 100 brisk (standard rating) 125 very fast 150 exceptionally fast The basic time for a task, or element, is the time for carrying out an element of work or an operation at standard rating.

Basic Time = Observed Time x Observed Rating

The result is expressed in basic minutes - BM's.

The work content of a job or operation is defined as: basic time + relaxation allowance + any allowance for additional work - e.g. that part of contingency allowance which represents work.

1.3 Standard Time

Standard time is the total time in which a job should be completed at standard performance i.e. work content, contingency allowance for delay, unoccupied time and interference allowance, where applicable.

Allowance for unoccupied time and for interference may be important for the measurement of machine-controlled operations, but they do not always appear in every computation of standard time. Relaxation allowance, on the other hand, has to be taken into account in every computation, whether the job is a simple manual one or a very complex operation requiring the simultaneous control of several machines. A contingency allowance will probably figure quite frequently in the compilation of standard times; it is therefore convenient to consider the contingency allowance and relaxation allowance, so that the sequence of calculation which started with the completion of observations at the workplace may be taken right through to the compilation of standard time.

Contingency allowance

A contingency allowance is a small allowance of time which may be included in a standard time to meet legitimate and expected items of work or delays, the precise measurement of which is uneconomical because of their infrequent or irregular occurrence.

Relaxation allowance

A relaxation allowance is an addition to the basic time to provide the worker with the opportunity to recover from physiological and psychological effects of carrying out specified work under specified conditions and to allow attention to personal needs. The amount of the allowance will depend on the nature of the job. Examples are:

Personal 5-7% Energy output 0-10% Noisy 0-5% Conditions 0-100% e.g. Electronics 5%

Other allowances

Other allowances include process allowance which is to cover when an operator is prevented from continuing with their work, although ready and waiting, by the process or machine requiring further time to complete its part of the job. A final allowance is that of Interference which is included whenever an operator has charge of more than one machine and the machines are subject to random stoppage. In normal circumstances the operator can only attend to one machine, and the others must wait for attention. This machine is then subject to interference which increased the machine cycle time.

It is now possible to obtain a complete picture of the standard time for a straightforward manual operation.

2. Activity Sampling

Activity sampling is a technique in which a large number of instantaneous observations are made over a period of time of a group of machines, processes or workers. Each observation records what is happening at that instant and the percentage of observations recorded for a particular activity or delay is a measure of the percentage of time during which the activity or delay occurs.

The advantages of this method are that

It is capable of measuring many activities that are impractical or too costly to be measured by time study. One observer can collect data concerning the simultaneous activities of a group. Activity sampling can be interrupted at any time without effect. The disadvantages are that

It is quicker and cheaper to use time study on jobs of short duration. It does not provide elemental detail. The type of information provided by an activity sampling study is:

a. The proportion of the working day during which workers or machines are producing.

b. The proportion of the working day used up by delays. The reason for each delay must be recorded.

c. The relative activity of different workers and machines.

To determine the number of observations in a full study the following equation is used:

Where:

3. Predetermined Motion Time Systems

A predetermined motion time system is a work measurement technique whereby times established for basic human motions (classified according to the nature of the motion and the conditions under which it is made) are used to build up the time for a job at a defined level of performance.

The systems are based on the assumption that all manual tasks can be analysed into basic motions of the body or body members. They were compiled as a result of a very large number of studies of each movement, generally by a frame-by-frame analysis of films of a wide range of subjects, men and women, performing a wide variety of tasks.

The first generation of PMT systems, MTM1, were very finely detailed, involving much analysis and producing extremely accurate results. This attention to detail was both a strength and a weakness, and for many potential applications the quantity of detailed analysis was not necessary, and prohibitively time -consuming. In these cases "second generation" techniques, such as Simplified PMTS, Master Standard Data, Primary Standard Data and MTM2, could be used with advantage, and no great loss of accuracy. For even speedier application, where some detail could be sacrificed then a "third generation" technique such as Basic Work Data or MTM3 could be used.

4. Synthesis

Synthesis is a work measurement technique for building up the time for a job at a defined level of performance by totaling element times obtained previously from time studies on other jobs containing the elements concerned, or from synthetic data.

Synthetic data is the name given to tables and formulae derived from the analysis of accumulated work measurement data, arranged in a form suitable for building up standard times, machine process times, etc. by synthesis.

Synthetic times are increasingly being used as a substitute for individual time studies in the case of jobs made up of elements which have recurred a sufficient number of times in jobs previously studied to make it possible to compile accurate representative times for them.

5. Estimating

The technique of estimating is the least refined of all those available to the work measurement practitioner. It consists of an estimate of total job duration (or in common practice, the job price or cost). This estimate is made by a craftsman or person familiar with the craft. It normally embraces the total components of the job, including work content, preparation and disposal time, any contingencies etc., all estimated in one gross amount.

6. Analytical estimating

This technique introduces work measurement concepts into estimating. In analytical estimating the estimator is trained in elemental breakdown, and in the concept of standard performance. The estimate is prepared by first breaking the work content of the job into elements, and then utilising the experience of the estimator (normally a craftsman) the time for each element of work is estimated - at standard performance. These estimated basic minutes are totalled to give a total job time, in basic minutes. An allowance for relaxation and any necessary contingency is then made, as in conventional time study, to give the standard time.

7. Comparative estimating

This technique has been developed to permit speedy and reliable assessment of the duration of variable and infrequent jobs, by estimating them within chosen time bands. Limits are set within which the job under consideration will fall, rather than in terms of precise capital standard or capital allowed minute values. It is applied by comparing the job to be estimated with jobs of similar work content, and using these similar jobs as "bench marks" to locate the new job in its relevant time band - known as Work Group.

Uses of Work Measurement

- To compare the efficiency of alternative methods. Other conditions being equal, the method which takes the least time will be the best method.

- To balance the work of members of teams, in association with the multiple activity charts, so that, as far as possible, each member has tasks taking an equal time.

- To determine, in association with man and machine multiple activity charts, the number of machines a worker can run.

Work Sampling

Work sampling is the statistical technique for determining the proportion of time spent by workers in various defined categories of activity (e.g. setting up a machine, assembling two parts, idle...etc.). It is as important as all other statistical techniques because it permits quick analysis, recognition, and enhancement of job responsibilities, tasks, performance competencies, and organizational work flows. Other names used for it are 'activity sampling', 'occurrence sampling', and 'ratio delay study'.

In a work sampling study, a large number of observations are made of the workers over an extended period of time. For statistical accuracy, the observations must be taken at random times during the period of study, and the period must be representative of the types of activities performed by the subjects.

One important usage of the work sampling technique is the determination of the standard time for a manual manufacturing task. Similar techniques for calculating the standard time are time study, standard data, and predetermined motion time systems.

Characteristics of Work Sampling Study

The study of work sampling has some general characteristics related to the work condition:

- One of them is the sufficient time available to perform the study. A work sampling study usually requires a substantial period of time to complete. There must be enough time available (several weeks or more) to conduct the study.

- Another characteristic is multiple workers. Work sampling is commonly used to study the activities of multiple workers rather than one worker.

- The third characteristic is long cycle time. The job covered in the study has relatively a long cycle time.

- The last condition is the non-repetitive work cycles. The work is not highly repetitive. The jobs consist of various tasks rather than a single repetitive task. However, it must be possible to classify the work activities into a distinct number of categories.

Steps in Conducting a Work Sampling Study

There are several recommended steps when starting to prepare a work sampling study:

1. Define the manufacturing tasks for which the standard time is to be determined.

2. Define the task elements. These are the defined broken-down steps of the task that will be observed during the study. Since a worker is going to be observed, additional categories will likely be included as well, such as "idle", "waiting for work", and "absent".

3. Design the study. This includes designing the forms that will be used to record the observations, determining how many observations will be required, deciding on the number of days or shifts to be included in the study, scheduling the observations, and finally determining the number of observers needed.

4. Identify the observers who will do the sampling.

5. Start the study. All those who are affected by the study should be informed about it.

6. Make random visits to the plant and collect the observations.

7. After completing the study, analyze and present the results. This is done by preparing a report that summarizes and analyzes all data and making recommendations when required.

Determining the Number of Observations Needed in Work Sampling

After the work elements are defined, the number of observations for the desired accuracy at the desired confidence level must be determined. The formula used in this method is:

$$\sigma_P = \sqrt{\frac{pq}{n}}$$

$$n = \frac{pq}{\sigma_P^{\,2}}$$

σ_P = standard error of proportion

p = percentage of idle time

q = percentage of working time

n = number of observations

Additional Applications of Work Sampling

Work sampling was initially developed for determining time allocation among workers' tasks in manufacturing environments. However, the technique has also been applied more broadly to examine work in a number of different environments, such as healthcare and construction. More recently, in the academic fields of organizational psychology and organizational behaviour, the basic technique has been developed into a detailed job analysis method for examining a range of different research questions.

Predetermined Motion Time System

A predetermined motion time system (PMTS) is frequently used to set labor rates in labour oriented industry by quantifying the amount of time required to perform specific tasks. The first such system is known as Methods-time measurement, released in 1948 and today existing in several variations, commonly known as MTM-1, MTM-2, MTM-UAS, MTM-MEK and MTM-B. Obsolete MTM standards include MTM-3 and MMMM (4M). The MTM-2 standard has also largely been phased out by the organization, but is still used in some commercial applications. Predetermined motion time system is another term to describe Predetermined Time standards (PTS).

MOST

Another popular PMTS is the Maynard Operation Sequence Technique, which was first released in 1972. H.B. Maynard was acquired by Accenture in 2007. That method also has several variations, with the most commonly used being BasicMOST, and others being MiniMOST, MaxiMOST, and AdminMOST. The variations of both systems differ from each other based on their level of focus. MTM-1 and MiniMOST are optimal for short processes with only small hand motions. BasicMOST, MODAPTS and MTM-UAS are more suitable for processes that average around 1 to 5 minutes, while MTM-B and MaxiMOST are more properly used for longer processes that are less repetitive.

MODAPTS

Yet another popular PMTS used today in the automotive, sewing and healthcare industries is the MODAPTS technique. This technique was introduced in 1966 by G. C. "Chris" Heyde who originally learned the MTM-1 and MTM-2 methods in the 1950s and sought a simpler technique to use and apply. Unlike the MTM and MOST standards, MODAPTS uses a MOD as its basic unit of measurement (1 MOD = 0.129 seconds). However, like Basic-MOST, MODAPTS uses a coding technique that consists of a letter and an integer number (all but 1 code), where the integer numbers each represent MODS that can be easily added to determine a coded task's time.

MODSEW

MODSEW is a software application of MODAPTS for the sewn products and apparel industries. It uses very intuitive codes to represent the various motion patterns prevalent in the industry and allows the user to configure their own codes for those unique to their operation. The software is used to determine the standard time to complete an operation and has provisions to collect and maintain groups of operations in a style (product). MODSEW is owned, maintained, sold and supported by Byte Software Services, LLC of Mauldin South Carolina.

GSD

General Sewing Data is a PMTS for the sewn products and apparel industries and is based on MTM Core Data both proprietary data systems of GSD (Corporate) Ltd of Preston, UK. The Time standards for General Sewing Data are used in GSD Enterprise and GSD QUEST. GSD company was taken over by thread giant Coats in 2015.

Pro SMV

Pro SMV is another PMTS for the sewn product industry, offered by Methods Apparel Consultancy India Private ltd. as an individual module of their software series pack called Pro-Suite. Pro Suite is aimed at providing various IT solutions for various departments of Sewn product industry. It has made a very strong consumer base in India and neighboring countries.

Seweasy

Seweasy is a more recent system used by Fortune 500 apparel brands, sourcing companies and SME sector manufacturers alike in Ready Made Garment (RMG) supply chain. Seweasy is more aligned with the lean concepts attributed to Toyota. This system focuses on providing Standard Minute Values (SAM, SMV) quickly for labour cost estimation in sewing. The living wage and escalating costs, "demand innovations and smooth work flow" in manufacturing location, to stay competitive and sustainable. The company provide free training on-site.

The Seweasy transparent garment sewing data is useful for "sewing load balancing" in line with Value Stream Mapping (VSM) and "added value" measurement known as "needle down time" among professionals. SewEasy Pvt. Ltd has trained many juniors and seniors alike, to quickly establish standards using this easy PMTS system. The process of Seweasy based sewing standardization leads to easier skills development and worker empowerment useful in performing innovations at needle point.

Transparent Garment Costing for Sourcing Teams

Recent research by Manchester University, UK on Garment Labor Costs and demand for living-wage benchmark brought Seweasy transparent garment sewing data and Methods-time measurement (MTM) to the notice of apparel industry's sourcing professionals, including Walmart, who adopted Seweasy to create a sustainable labour-costing model for sourcing garments and home textiles in 2013. This approach is explained by ASDA in Clean Clothes Campaign annual report too at https://www.cleanclothes.org/livingwage/tailoredwages/tailored-wages-position/asdaprofile.pdf.

Home Textiles, Shoes & Leather

Seweasy can be used in all needle based sewing industries such as Home Textiles, shoe, leather, and upholstery.

Seweasy Lean Approach for Muda, Mura and Muri

Unlike 'time studies', in which an analyst uses a stopwatch and subjectively rates the operator's effort to calculate a standard time, Seweasy PMTS requires that the analyst break apart the process into its elemental sewing actions, and add-up the times to calculate the total standard time. Process of estimating cost of labour and workflow, professionally in such manner leads to improved productivity, lowered cost and higher earnings for all stake holders by way of eliminating wastes Muda, Mura and Muri.

TMU

Most predetermined motion time systems (MTM and MOST) use time measurement units (TMU) instead of seconds for measuring time. One TMU is defined to be 0.00001 hours, or 0.036 seconds. These smaller units allow for more accurate calculations without the use of decimals. In the most in-depth PMT systems, motions observed will be on the level of individual TMUs, like toss (3 TMUs in MiniMOST) and simple pick-up (2 TMUs in MTM-1). More general systems simplify things by grouping individual elements, and thus have larger time values – for example, a bend and arise (61 TMUs in MTM-2) and one or two steps (30 TMUs in BasicMOST). Systems with even less detail work with TMU values in the hundreds, like climbing 10 rungs on a ladder (300 TMUs in MaxiMOST) or passing through a door (100 TMUs in MaxiMOST).

The choice of which variation of a certain PMTS to use is dependent on the need for accuracy in contrast to the need for quick analysis, as well as the length of the operation, the distances involved in the operation, and the repetitiveness of the operation. Longer operations often take place on a larger spatial scale, and tend to be less repetitive, so these issues are often treated as one. For longer, less repetitive operations, statistical analysis demonstrates that the accuracy of less detailed systems will generally approach the accuracy of more detailed systems. Thus, in order to reduce the time required for analysis, less detailed systems (like MTM-B and MaxiMOST) are usually used when possible. Conversely, very short, repetitive processes are commonly analyzed with more exact methods like MTM-1 and MiniMOST because of the need for accuracy.

Ilo & Manchester University Research During 2009 to 2013

Manchester University researchers including Doug Miller, has gone deep in to uses of PMTS in apparel labour costing in "Towards Sustainable Labour Costing in UK Fashion Retail." Doug says ..work measurement for arriving at a standard time should normally make provision for relax-ation, contingency and special allowances. According to the International Labour Organization (ILO), as of 1992 there were some 200 different PTS systems. In apparel manufacture, three PTS consultancy firms specializing in MTM appear to be operating in the sector– the US-based MODAPTS, the Sri Lankan-based Seweasy and the UK-headquartered GSD (Corporate) Ltd.

Comparison of Pmts Systems by Walmart

Walmart has launched in house PMTS system under lean cost reduction program across Asia.

Automation

Automation or automatic control, is the use of various control systems for operating equipment such as machinery, processes in factories, boilers and heat treating ovens, switching on telephone networks, steering and stabilization of ships, aircraft and other applications and vehicles with minimal or reduced human intervention. Some processes have been completely automated.

The biggest benefit of automation is that it saves labor; however, it is also used to save energy and materials and to improve quality, accuracy and precision.

The term *automation*, inspired by the earlier word *automatic* (coming from *automaton*), was not widely used before 1947, when Ford established an automation department. It was during this time that industry was rapidly adopting feedback controllers, which were introduced in the 1930s.

Automation has been achieved by various means including mechanical, hydraulic, pneumatic, electrical, electronic devices and computers, usually in combination. Complicated systems, such as modern factories, airplanes and ships typically use all these combined techniques.

Open-Loop and Closed-Loop (Feedback) Control

Fundamentally, there are two types of control loop; open loop control, and closed loop (feedback) control.

In open loop control, the control action from the controller is independent of the "process output" (or "controlled process variable"). A good example of this is a central heating boiler controlled only by a timer, so that heat is applied for a constant time, regardless of the temperature of the building. (The control action is the switching on/off of the boiler. The process output is the building temperature).

In closed loop control, the control action from the controller is dependent on the process output. In the case of the boiler analogy this would include a thermostat to monitor the building temperature, and thereby feed back a signal to ensure the controller maintains the building at the temperature set on the thermostat. A closed loop controller therefore has a feedback loop which ensures the controller exerts a control action to give a process output the same as the "Reference input" or "set point". For this reason, closed loop controllers are also called feedback controllers.

The definition of a closed loop control system according to the British Standard Institution is 'a control system possessing monitoring feedback, the deviation signal formed as a result of this feedback being used to control the action of a final control element in such a way as to tend to reduce the deviation to zero.' "

Likewise; "A *Feedback Control System* is a system which tends to maintain a prescribed relationship of one system variable to another by comparing functions of these variables and using the difference as a means of control.'"

The advanced type of automation that revolutionized manufacturing, aircraft, communications and other industries, is feedback control, which is usually *continuous* and involves taking measurements using a sensor and making calculated adjustments to keep the measured variable within a set range. The theoretical basis of closed loop automation is control theory.

Control Actions

Discrete Control (On/Off)

One of the simplest types of control is *on-off* control. An example is the thermostat used on household appliances which either opens or closes an electrical contact. (Thermostats were originally developed as true feedback-control mechanisms rather than the on-off common household appliance thermostat.)

Sequence control, in which a programmed sequence of *discrete* operations is performed, often based on system logic that involves system states. An elevator control system is an example of sequence control.

PID Controller

A proportional–integral–derivative controller (PID controller) is a control loop feedback mechanism (controller) widely used in industrial control systems.

A PID controller continuously calculates an *error value* e (t) as the difference between a desired setpoint and a measured process variable and applies a correction based on proportional, integral, and derivative terms, respectively (sometimes denoted *P*, *I*, and *D*) which give their name to the controller type.

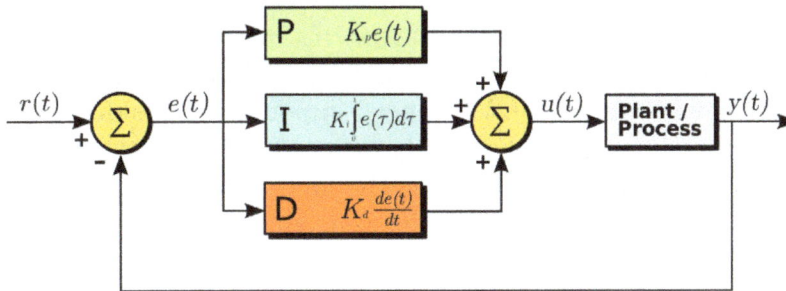

A block diagram of a PID controller in a feedback loop, r(t) is the desired process value or "set point", and y(t) is the measured process value.

The theoretical understanding and application dates from the 1920s, and they are implemented in nearly all analogue control systems; originally in mechanical controllers, and then using discrete electronics and latterly in industrial process computers.

Sequential Control and Logical Sequence or System State Control

Sequential control may be either to a fixed sequence or to a logical one that will perform different actions depending on various system states. An example of an adjustable but otherwise fixed sequence is a timer on a lawn sprinkler.

States refer to the various conditions that can occur in a use or sequence scenario of the system. An example is an elevator, which uses logic based on the system state to perform certain actions in response to its state and operator input. For example, if the operator presses the floor n button, the system will respond depending on whether the elevator is stopped or moving, going up or down, or if the door is open or closed, and other conditions.

An early development of sequential control was relay logic, by which electrical relays engage electrical contacts which either start or interrupt power to a device. Relays were first used in telegraph networks before being developed for controlling other devices, such as when starting and stopping industrial-sized electric motors or opening and closing solenoid valves. Using relays for control purposes allowed event-driven control, where actions could be triggered out of sequence, in response to external events. These were more flexible in their response than the rigid single-sequence cam timers. More complicated examples involved maintaining safe sequences for devices such as swing bridge controls, where a lock bolt needed to be disengaged before the bridge could be moved, and the lock bolt could not be released until the safety gates had already been closed.

The total number of relays, cam timers and drum sequencers can number into the hundreds or even thousands in some factories. Early programming techniques and languages were needed to make such systems manageable, one of the first being ladder logic, where diagrams of the interconnected relays resembled the rungs of a ladder. Special computers called programmable logic controllers were later designed to replace these collections of hardware with a single, more easily re-programmed unit.

In a typical hard wired motor start and stop circuit (called a *control circuit*) a motor is started by pushing a "Start" or "Run" button that activates a pair of electrical relays. The "lock-in" relay locks in contacts that keep the control circuit energized when the push button is released. (The start button is a normally open contact and the stop button is normally closed contact.) Another relay energizes a switch that powers the device that throws the motor starter switch (three sets of contacts for three phase industrial power) in the main power circuit. Large motors use high voltage and experience high in-rush current, making speed important in making and breaking contact. This can be dangerous for personnel and property with manual switches. The "lock in" contacts in the start circuit and the main power contacts for the motor are held engaged by their respective electromagnets until a "stop" or "off" button is pressed, which de-energizes the lock in relay.

Commonly interlocks are added to a control circuit. Suppose that the motor in the example is powering machinery that has a critical need for lubrication. In this case an interlock could be added to insure that the oil pump is running before the motor starts. Timers, limit switches and electric eyes are other common elements in control circuits.

Solenoid valves are widely used on compressed air or hydraulic fluid for powering actuators on mechanical components. While motors are used to supply continuous rotary motion, actuators are typically a better choice for intermittently creating a limited range of movement for a mechanical component, such as moving various mechanical arms, opening or closing valves, raising heavy press rolls, applying pressure to presses.

Computer Control

Computers can perform both sequential control and feedback control, and typically a single computer will do both in an industrial application. Programmable logic controllers (PLCs) are a type of special purpose microprocessor that replaced many hardware components such as timers and drum sequencers used in relay logic type systems. General purpose process control computers have increasingly replaced stand alone controllers, with a single computer able to perform the operations of hundreds of controllers. Process control computers can process data from a network of PLCs, instruments and controllers in order to implement typical (such as PID) control of many individual variables or, in some cases, to implement complex control algorithms using multiple inputs and mathematical manipulations. They can also analyze data and create real time graphical displays for operators and run reports for operators, engineers and management.

Control of an automated teller machine (ATM) is an example of an interactive process in which a computer will perform a logic derived response to a user selection based on information retrieved from a networked database. The ATM process has similarities with other online transaction processes. The different logical responses are called *scenarios*. Such processes are typically designed with the aid of use cases and flowcharts, which guide the writing of the software code.

History

The earliest feedback control mechanism was the thermostat invented in 1620 by the Dutch scientist Cornelius Drebbel. (Note: Early thermostats were temperature regulators or controlers rather than the on-off mechanisms common in household appliances.) Another control mechanism was used to tent the sails of windmills. It was patented by Edmund Lee in 1745. Also in 1745, Jacques de Vaucanson invented the first automated loom.

In 1771 Richard Arkwright invented the first fully automated spinning mill driven by water power, known at the time as the water frame.

The centrifugal governor, which was invented by Christian Huygens in the seventeenth century, was used to adjust the gap between millstones. The centrifugal governor was also used in the automatic flour mill developed by Oliver Evans in 1785, making it the first completely automated industrial process. The governor was adopted by James Watt for use on a steam engine in 1788 after Watt's partner Boulton saw one at a flour mill Boulton & Watt were building.

The governor could not actually hold a set speed; the engine would assume a new constant speed in response to load changes. The governor was able to handle smaller variations such as those caused by fluctuating heat load to the boiler. Also, there was a tendency for oscillation whenever there was a speed change. As a consequence, engines equipped with this governor were not suitable for operations requiring constant speed, such as cotton spinning.

Several improvements to the governor, plus improvements to valve cut-off timing on the steam engine, made the engine suitable for most industrial uses before the end of the 19th century. Advances in the steam engine stayed well ahead of science, both thermodynamics and control theory.

The governor received relatively little scientific attention until James Clerk Maxwell published a paper that established the beginning of a theoretical basis for understanding control theory. Development of the electronic amplifier during the 1920s, which was important for long distance telephony, required a higher signal to noise ratio, which was solved by negative feedback noise cancellation. This and other telephony applications contributed to control theory. Military applications during the Second World War that contributed to and benefited from control theory were fire-control systems and aircraft controls. The word "automation" itself was coined in the 1940s by General Electric. The so-called classical theoretical treatment of control theory dates to the 1940s and 1950s.

Relay logic was introduced with factory electrification, which underwent rapid adaption from 1900 though the 1920s. Central electric power stations were also undergoing rapid growth and operation of new high pressure boilers, steam turbines and electrical substations created a large demand for instruments and controls.

Central control rooms became common in the 1920s, but as late as the early 1930s, most process control was on-off. Operators typically monitored charts drawn by recorders that plotted data from instruments. To make corrections, operators manually opened or closed valves or turned switches on or off. Control rooms also used color coded lights to send signals to workers in the plant to manually make certain changes.

Controllers, which were able to make calculated changes in response to deviations from a set point rather than on-off control, began being introduced the 1930s. Controllers allowed manufacturing to continue showing productivity gains to offset the declining influence of factory electrification.

Factory productivity was greatly increased by electrification in the 1920s. Manufacturing productivity growth fell from 5.2%/yr 1919-29 to 2.76%/yr 1929-41. Field notes that spending on non-medical instruments increased significantly from 1929–33 and remained strong thereafter.

In 1959 Texaco's Port Arthur refinery became the first chemical plant to use digital control. Conversion of factories to digital control began to spread rapidly in the 1970s as the price of computer hardware fell.

Significant Applications

The automatic telephone switchboard was introduced in 1892 along with dial telephones. By 1929, 31.9% of the Bell system was automatic. Automatic telephone switching originally used vacuum tube amplifiers and electro-mechanical switches, which consumed a large amount of electricity. Call volume eventually grew so fast that it was feared the telephone system would consume all electricity production, prompting Bell Labs to begin research on the transistor.

The logic performed by telephone switching relays was the inspiration for the digital computer. The first commercially successful glass bottle blowing machine was an automatic model introduced in 1905. The machine, operated by a two-man crew working 12-hour shifts, could produce 17,280 bottles in 24 hours, compared to 2,880 bottles made by a crew of six men and boys working in a shop for a day. The cost of making bottles by machine was 10 to 12 cents per gross compared to $1.80 per gross by the manual glassblowers and helpers.

Sectional electric drives were developed using control theory. Sectional electric drives are used on different sections of a machine where a precise differential must be maintained between the sections. In steel rolling, the metal elongates as it passes through pairs of rollers, which must run at successively faster speeds. In paper making the paper sheet shrinks as it passes around steam heated drying arranged in groups, which must run at successively slower speeds. The first application of a sectional electric drive was on a paper machine in 1919. One of the most important developments in the steel industry during the 20th century was continuous wide strip rolling, developed by Armco in 1928.

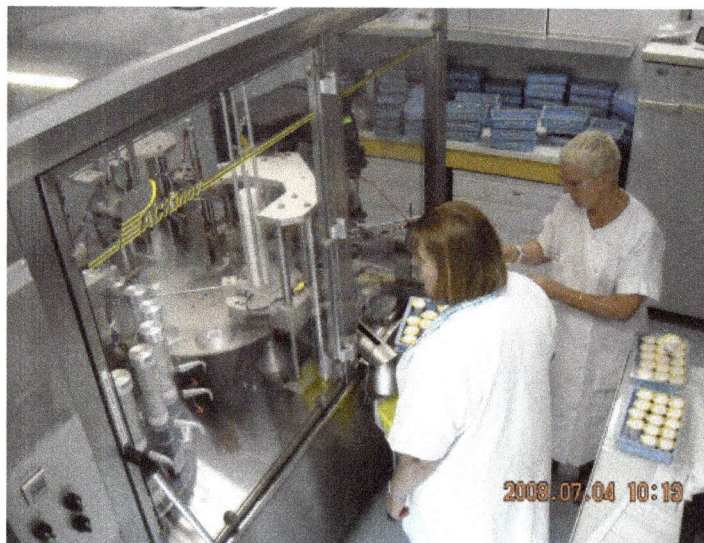

Automated pharmacology production

Before automation many chemicals were made in batches. In 1930, with the widespread use of instruments and the emerging use of controllers, the founder of Dow Chemical Co. was advocating continuous production.

Self-acting machine tools that displaced hand dexterity so they could be operated by boys and unskilled laborers were developed by James Nasmyth in the 1840s. Machine tools were automated with Numerical control (NC) using punched paper tape in the 1950s. This soon evolved into computerized numerical control (CNC).

Today extensive automation is practiced in practically every type of manufacturing and assembly process. Some of the larger processes include electrical power generation, oil refining, chemicals, steel mills, plastics, cement plants, fertilizer plants, pulp and paper mills, automobile and truck assembly, aircraft production, glass manufacturing, natural gas separation plants, food and beverage processing, canning and bottling and manufacture of various kinds of parts. Robots are especially useful in hazardous applications like automobile spray painting. Robots are also used to assemble electronic circuit boards. Automotive welding is done with robots and automatic welders are used in applications like pipelines.

Advantages and Disadvantages

The main advantages of automation are:

- Increased throughput or productivity.
- Improved quality or increased predictability of quality.
- Improved robustness (consistency), of processes or product.
- Increased consistency of output.
- Reduced direct human labor costs and expenses.

The following methods are often employed to improve productivity, quality, or robustness.

- Install automation in operations to reduce cycle time.
- Install automation where a high degree of accuracy is required.
- Replacing human operators in tasks that involve hard physical or monotonous work.
- Replacing humans in tasks done in dangerous environments (i.e. fire, space, volcanoes, nuclear facilities, underwater, etc.)
- Performing tasks that are beyond human capabilities of size, weight, speed, endurance, etc.
- Reduces operation time and work handling time significantly.
- Frees up workers to take on other roles.
- Provides higher level jobs in the development, deployment, maintenance and running of the automated processes.

The main disadvantages of automation are:

- Security Threats/Vulnerability: An automated system may have a limited level of intelligence, and is therefore more susceptible to committing errors outside of its immediate scope of knowledge (e.g., it is typically unable to apply the rules of simple logic to general propositions).

- Unpredictable/excessive development costs: The research and development cost of automating a process may exceed the cost saved by the automation itself.

- High initial cost: The automation of a new product or plant typically requires a very large initial investment in comparison with the unit cost of the product, although the cost of automation may be spread among many products and over time.

In manufacturing, the purpose of automation has shifted to issues broader than productivity, cost, and time.

Lights Out Manufacturing

Lights out manufacturing is when a production system is 100% or near to 100% automated (not hiring any workers). In order to eliminate the need for labor costs altogether.

Health and Environment

The costs of automation to the environment are different depending on the technology, product or engine automated. There are automated engines that consume more energy resources from the Earth in comparison with previous engines and those that do the opposite too. Hazardous operations, such as oil refining, the manufacturing of industrial chemicals, and all forms of metal working, were always early contenders for automation.

Convertibility and Turnaround Time

Another major shift in automation is the increased demand for flexibility and convertibility in manufacturing processes. Manufacturers are increasingly demanding the ability to easily switch from manufacturing Product A to manufacturing Product B without having to completely rebuild the production lines. Flexibility and distributed processes have led to the introduction of Automated Guided Vehicles with Natural Features Navigation.

Digital electronics helped too. Former analogue-based instrumentation was replaced by digital equivalents which can be more accurate and flexible, and offer greater scope for more sophisticated configuration, parametrization and operation. This was accompanied by the fieldbus revolution which provided a networked (i.e. a single cable) means of communicating between control systems and field level instrumentation, eliminating hard-wiring.

Discrete manufacturing plants adopted these technologies fast. The more conservative process industries with their longer plant life cycles have been slower to adopt and analogue-based measurement and control still dominates. The growing use of Industrial Ethernet on the factory floor is pushing these trends still further, enabling manufacturing plants to be integrated more tightly within the enterprise, via the internet if necessary. Global competition has also increased demand for Reconfigurable Manufacturing Systems.

Automation Tools

Engineers can now have numerical control over automated devices. The result has been a rapidly expanding range of applications and human activities. Computer-aided technologies (or CAx) now serve as the basis for mathematical and organizational tools used to create complex systems. Notable examples of CAx include Computer-aided design (CAD software) and Computer-aided manufacturing (CAM software). The improved design, analysis, and manufacture of products enabled by CAx has been beneficial for industry.

Information technology, together with industrial machinery and processes, can assist in the design, implementation, and monitoring of control systems. One example of an industrial control system is a programmable logic controller (PLC). PLCs are specialized hardened computers which are frequently used to synchronize the flow of inputs from (physical) sensors and events with the flow of outputs to actuators and events.

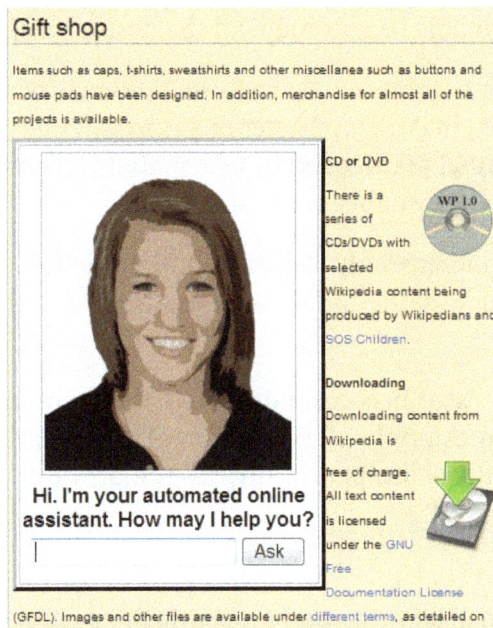

An automated online assistant on a website, with an avatar for enhanced human–computer interaction.

Human-machine interfaces (HMI) or computer human interfaces (CHI), formerly known as *man-machine interfaces*, are usually employed to communicate with PLCs and other computers. Service personnel who monitor and control through HMIs can be called by different names. In industrial process and manufacturing environments, they are called operators or something similar. In boiler houses and central utilities departments they are called stationary engineers.

Different types of automation tools exist:

- ANN - Artificial neural network
- DCS - Distributed Control System
- HMI - Human Machine Interface
- SCADA - Supervisory Control and Data Acquisition
- PLC - Programmable Logic Controller
- Instrumentation

- Motion control

- Robotics

When it comes to Factory Automation, Host Simulation Software (HSS) is a commonly used testing tool that is used to test the equipment software. HSS is used to test equipment performance with respect to Factory Automation standards (timeouts, response time, processing time).

Limitations to Automation

- Current technology is unable to automate all the desired tasks.

- Many operations using automation have large amounts of invested capital and produce high volumes of product, making malfunctions extremely costly and potentially hazardous. Therefore, some personnel are needed to insure that the entire system functions properly and that safety and product quality are maintained.

- As a process becomes increasingly automated, there is less and less labor to be saved or quality improvement to be gained. This is an example of both diminishing returns and the logistic function.

- As more and more processes become automated, there are fewer remaining non-automated processes. This is an example of exhaustion of opportunities. New technological paradigms may however set new limits that surpass the previous limits.

Current Limitations

Many roles for humans in industrial processes presently lie beyond the scope of automation. Human-level pattern recognition, language comprehension, and language production ability are well beyond the capabilities of modern mechanical and computer systems. Tasks requiring subjective assessment or synthesis of complex sensory data, such as scents and sounds, as well as high-level tasks such as strategic planning, currently require human expertise. In many cases, the use of humans is more cost-effective than mechanical approaches even where automation of industrial tasks is possible. Overcoming these obstacles is a theorized path to post-scarcity economics.

Paradox of Automation

The Paradox of Automation says that the more efficient the automated system, the more crucial the human contribution of the operators. Humans are less involved, but their involvement becomes more critical.

If an automated system has an error, it will multiply that error until it's fixed or shut down. This is where human operators come in.

A fatal example of this was Air France Flight 447, where a failure of automation put the pilots into a manual situation they were not prepared for.

Recent and Emerging Applications

KUKA industrial robots being used at a bakery for food production

Automated Retail

Food and Drink

The food retail industry has started to apply automation to the ordering process; McDonald's has introduced touch screen ordering and payment systems in many of its restaurants, reducing the need for as many cashier employees. The University of Texas at Austin has introduced fully automated cafe retail locations. Some Cafes and restaurants have utilized mobile and tablet "apps" to make the ordering process more efficient by customers ordering and paying on their device. Some restaurants have automated food delivery to customers tables using a Conveyor belt system. The use of robots is sometimes employed to replace waiting staff.

Stores

Many Supermarkets and even smaller stores are rapidly introducing Self checkout systems reducing the need for employing checkout workers.

Online shopping could be considered a form of automated retail as the payment and checkout are through an automated Online transaction processing system. Other forms of automation can also be an integral part of online shopping, for example the deployment of automated warehouse robotics such as that applied by Amazon using Kiva Systems.

Automated Mining

Involves the removal of human labor from the mining process. The mining industry is currently in the transition towards Automation. Currently it can still require a large amount of human capital, particularly in the third world where labor costs are low so there is less incentive for increasing efficiency through automation.

Automated Video Surveillance

The Defense Advanced Research Projects Agency (DARPA) started the research and development of automated visual surveillance and monitoring (VSAM) program, between 1997 and 1999, and airborne video surveillance (AVS) programs, from 1998 to 2002. Currently, there is a major effort underway in the vision community to develop a fully automated tracking surveillance system. Automated video surveillance monitors people and vehicles in real time within a busy environment. Existing automated surveillance systems are based on the environment they are primarily designed to observe, i.e., indoor, outdoor or airborne, the amount of sensors that the automated system can handle and the mobility of sensor, i.e., stationary camera vs. mobile camera. The purpose of a surveillance system is to record properties and trajectories of objects in a given area, generate warnings or notify designated authority in case of occurrence of particular events.

Automated Highway Systems

As demands for safety and mobility have grown and technological possibilities have multiplied, interest in automation has grown. Seeking to accelerate the development and introduction of fully automated vehicles and highways, the United States Congress authorized more than $650 million over six years for intelligent transport systems (ITS) and demonstration projects in the 1991 Intermodal Surface Transportation Efficiency Act (ISTEA). Congress legislated in ISTEA that "the Secretary of Transportation shall develop an automated highway and vehicle prototype from which future fully automated intelligent vehicle-highway systems can be developed. Such development shall include research in human factors to ensure the success of the man-machine relationship. The goal of this program is to have the first fully automated highway roadway or an automated test track in operation by 1997. This system shall accommodate installation of equipment in new and existing motor vehicles." [ISTEA 1991, part B, Section 6054(b)].

Full automation commonly defined as requiring no control or very limited control by the driver; such automation would be accomplished through a combination of sensor, computer, and communications systems in vehicles and along the roadway. Fully automated driving would, in theory, allow closer vehicle spacing and higher speeds, which could enhance traffic capacity in places where additional road building is physically impossible, politically unacceptable, or prohibitively expensive. Automated controls also might enhance road safety by reducing the opportunity for driver error, which causes a large share of motor vehicle crashes. Other potential benefits include improved air quality (as a result of more-efficient traffic flows), increased fuel economy, and spin-off technologies generated during research and development related to automated highway systems.

Automated Waste Management

Automated waste collection trucks prevent the need for as many workers as well as easing the level of labor required to provide the service.

Home Automation

Home automation (also called domotics) designates an emerging practice of increased automation of household appliances and features in residential dwellings, particularly through electronic means that allow for things impracticable, overly expensive or simply not possible in recent past decades.

Laboratory Automation

Automation is essential for many scientific and clinical applications. Therefore, automation has been extensively employed in laboratories. From as early as 1980 fully automated laboratories have already been working. However, automation has not become widespread in laboratories due to its high cost. This may change with the ability of integrating low-cost devices with standard laboratory equipment. Autosamplers are common devices used in laboratory automation.

Industrial Automation

Industrial automation deals primarily with the automation of manufacturing, quality control and material handling processes. General purpose controllers for industrial processes include Programmable logic controllers, stand-alone I/O modules, and computers. Industrial automation is to replace the decision making of humans and manual command-response activities with the use of mechanized equipment and logical programming commands. One trend is increased use of Machine vision to provide automatic inspection and robot guidance functions, another is a continuing increase in the use of robots. Industrial automation is simply done at the industrial level.

Energy efficiency in industrial processes has become a higher priority. Semiconductor companies like Infineon Technologies are offering 8-bit micro-controller applications for example found in motor controls, general purpose pumps, fans, and ebikes to reduce energy consumption and thus increase efficiency.

Advantages

- Replaces hard physical or monotonous work
- Tasks in hazardous environments, such as extreme temperatures, or atmospheres that are radioactive or toxic can be done by machines
- Faster production and cheaper labor costs
- Automation can be maintained with simple quality checks.
- Can perform tasks beyond human capabilities.

Disadvantages

- As of now, not all tasks can be automated
- Some tasks are more expensive to automate
- Initial costs are high
- Failure to maintain a system could result in the loss of the product

Industrial Robotics

Industrial robotics is a sub-branch in the industrial automation that aids in various manufacturing processes. Such manufacturing processes include; machining, welding, painting, assembling and material handling to name a few. Industrial robots utilizes various mechanical, electrical as well as software systems to allow for high precision, accuracy and speed that far exceeds any human performance. The birth of industrial robot came shortly after World War II as United States saw the need for a quicker way to produce industrial and consumer goods. Servos, digital logic and solid state electronics allowed engineers to build better and faster systems and overtime these systems were improved and revised to the point where a single robot is capable of running 24 hours a day with little or no maintenance.

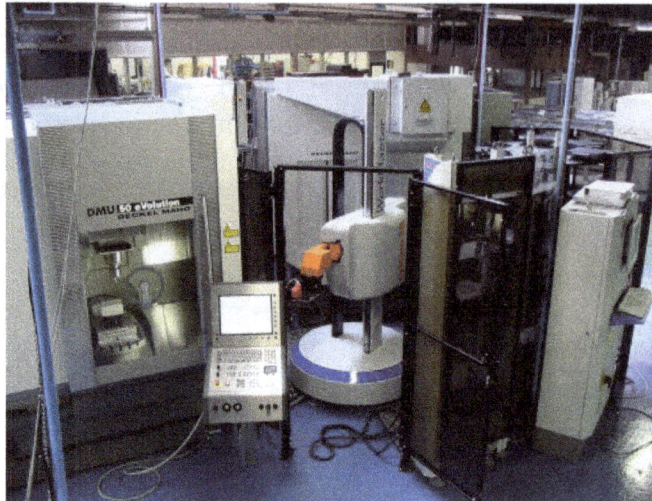
Automated milling machines

Programmable Logic Controllers

Industrial automation incorporates programmable logic controllers in the manufacturing process. Programmable logic controllers (PLCs) use a processing system which allows for variation of controls of inputs and outputs using simple programming. PLCs make use of programmable memory, storing instructions and functions like logic, sequencing, timing, counting, etc. Using a logic based language, a PLC can receive a variety of inputs and return a variety of logical outputs, the input devices being sensors and output devices being motors, valves, etc. PLCs are similar to computers, however, while computers are optimized for calculations, PLCs are optimized for control task and use in industrial environments. They are built so that only basic logic-based programming knowledge is needed and to handle vibrations, high temperatures, humidity and noise. The greatest advantage PLCs offer is their flexibility. With the same basic controllers, a PLC can operate a range of different control systems. PLCs make it unnecessary to rewire a system to change the control system. This flexibility leads to a cost-effective system for complex and varied control systems.

Agent-assisted Automation

Agent-assisted automation refers to automation used by call center agents to handle customer inquiries. There are two basic types: desktop automation and automated voice solutions. Desktop automation refers to software programming that makes it easier for the call center agent to work across multiple desktop tools. The automation would take the information entered into one tool and populate it across the others so it did not have to be entered more than once, for example. Automated voice solutions allow the agents to remain on the line while disclosures and other important information is provided to customers in the form of pre-recorded audio files. Specialized applications of these automated voice solutions enable the agents to process credit cards without ever seeing or hearing the credit card numbers or CVV codes

The key benefit of agent-assisted automation is compliance and error-proofing. Agents are sometimes not fully trained or they forget or ignore key steps in the process. The use of automation ensures that what is supposed to happen on the call actually does, every time.

Relationship to Unemployment

Research by the Oxford Martin School showed that employees engaged in "tasks following well-defined procedures that can easily be performed by sophisticated algorithms" are at risk of displacement. The study, published in 2013, shows that automation can affect both skilled and unskilled work and both high and low-paying occupations; however, low-paid physical occupations are most at risk. However, according to a study published in McKinsey Quarterly in 2015 the impact of computerization in most cases is not replacement of employees but automation of portions of the tasks they perform.

Based on a formula by Gilles Saint-Paul, an economist at Toulouse 1 University, the demand for unskilled human capital declines at a slower rate than the demand for skilled human capital increases. In the long run and for society as a whole it has led to cheaper products, lower average work hours, and new industries forming (I.e, robotics industries, computer industries, design industries). These new industries provide many high salary skill based jobs to the economy.

Casting

Molten metal before casting

Casting iron in a sand mold

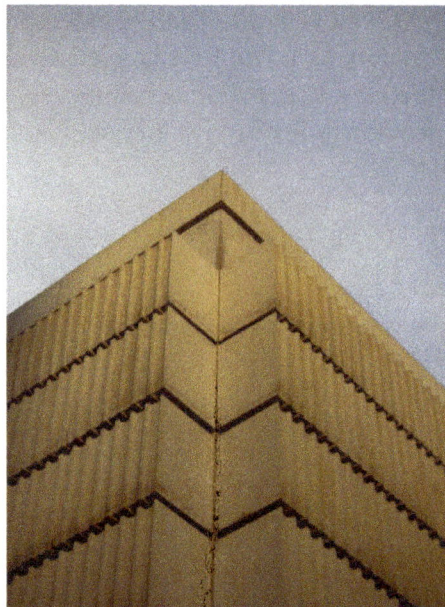

Judenplatz Holocaust Memorial (Nameless Library), by Rachel Whiteread.
Concrete cast of books on library shelves turned inside out.

Casting is a manufacturing process in which a liquid material is usually poured into a mold, which contains a hollow cavity of the desired shape, and then allowed to solidify. The solidified part is also known as a *casting*, which is ejected or broken out of the mold to complete the process. Casting materials are usually metals or various *cold setting* materials that cure after mixing two or more components together; examples are epoxy, concrete, plaster and clay. Casting is most often used for making complex shapes that would be otherwise difficult or uneconomical to make by other methods.

Casting is a 6000-year-old process. The oldest surviving casting is a copper frog from 3200 BC.

Types

Metal

In metalworking, metal is heated until it becomes liquid and is then poured into a mold. The mold is a hollow cavity that includes the desired shape, but the mold also includes runners and risers that enable the metal to fill the mold. The mold and the metal are then cooled until the metal solidifies. The solidified part (the casting) is then recovered from the mold. Subsequent operations remove excess material caused by the casting process (such as the runners and risers).

Plaster, Concrete, or Plastic Resin

Plaster and other chemical curing materials such as concrete and plastic resin may be cast using single-use *waste* molds as noted above, multiple-use 'piece' molds, or molds made of small rigid pieces or of flexible material such as latex rubber (which is in turn supported by an exterior mold). When casting plaster or concrete, the material surface is flat and lacks transparency. Often topical treatments are applied to the surface. For example, painting and etching can be used in a way that give the appearance of metal or stone. Alternatively, the material is altered in its initial casting process and may contain colored sand so as to give an appearance of stone. By casting concrete, rather than plaster, it is possible to create sculptures, fountains, or seating for outdoor use. A simulation of high-quality marble may be made using certain chemically-set plastic resins (for example epoxy or polyester) with powdered stone added for coloration, often with multiple colors worked in. The latter is a common means of making washstands, washstand tops and shower stalls, with the skilled working of multiple colors resulting in simulated staining patterns as is often found in natural marble or travertine.

Fettling

Raw castings often contain irregularities caused by seams and imperfections in the molds, as well as access ports for pouring material into the molds. The process of cutting, grinding, shaving or sanding away these unwanted bits is called "fettling". In modern times robotic processes have been developed to perform some of the more repetitive parts of the fettling process, but historically fettlers carried out this arduous work manually, and often in conditions dangerous to their health.

Fettling can add significantly to the cost of the resulting product, and designers of molds seek to minimize it through the shape of the mold, the material being cast, and sometimes by including decorative elements.

Casting Process Simulation

Casting process simulation uses numerical methods to calculate cast component quality considering mold filling, solidification and cooling, and provides a quantitative prediction of casting mechanical properties, thermal stresses and distortion. Simulation accurately describes a cast component's quality up-front before production starts. The casting rigging can be designed with respect to the required component properties. This has benefits beyond a reduction in pre-production sampling, as the precise layout of the complete casting system also leads to energy, material, and tooling savings.

The software supports the user in component design, the determination of melting practice and casting methoding through to pattern and mold making, heat treatment, and finishing. This saves costs along the entire casting manufacturing route.

Casting process simulation was initially developed at universities starting from the early '70s, mainly in Europe and in the U.S., and is regarded as the most important innovation in casting technology over the last 50 years. Since the late '80s, commercial programs (such as AutoCAST and MAGMA) are available which make it possible for foundries to gain new insight into what is happening inside the mold or die during the casting process.

Computer-Aided Manufacturing

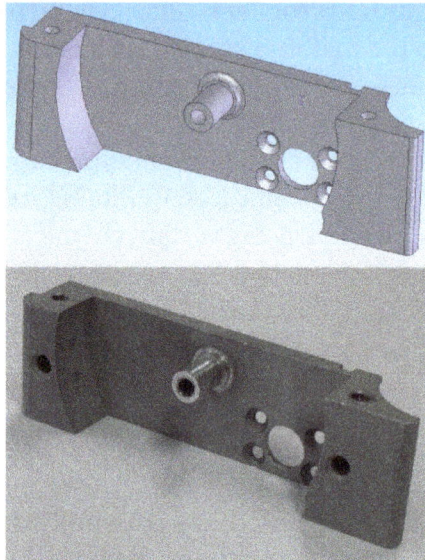

CAD model and CNC machined part

Computer-aided manufacturing (CAM) is the use of software to control machine tools and related ones in the manufacturing of workpieces. This is not the only definition for CAM, but it is the most common; CAM may also refer to the use of a computer to assist in all operations of a manufacturing plant, including planning, management, transportation and storage. Its primary purpose is to create a faster production process and components and tooling with more precise dimensions and material consistency, which in some cases, uses only the required amount of raw material (thus minimizing waste), while simultaneously reducing energy consumption. CAM is now a system used in schools and lower educational purposes. CAM is a subsequent computer-aided process after computer-aided design (CAD) and sometimes computer-aided engineering (CAE), as the model generated in CAD and verified in CAE can be input into CAM software, which then controls the machine tool. CAM is used in many schools alongside computer-aided design (CAD) to create objects.

Overview

Traditionally, CAM has been considered as a numerical control (NC) programming tool, where in two-dimensional (2-D) or three-dimensional (3-D) models of components generated in CADAs with other "Computer-Aided" technologies, CAM does not eliminate the need for skilled professionals such as manufacturing engineers, NC programmers, or machinists. CAM, in fact, leverages both the value of the most skilled manufacturing professionals through advanced productivity tools, while building the skills of new professionals through visualization, simulation and optimization tools.

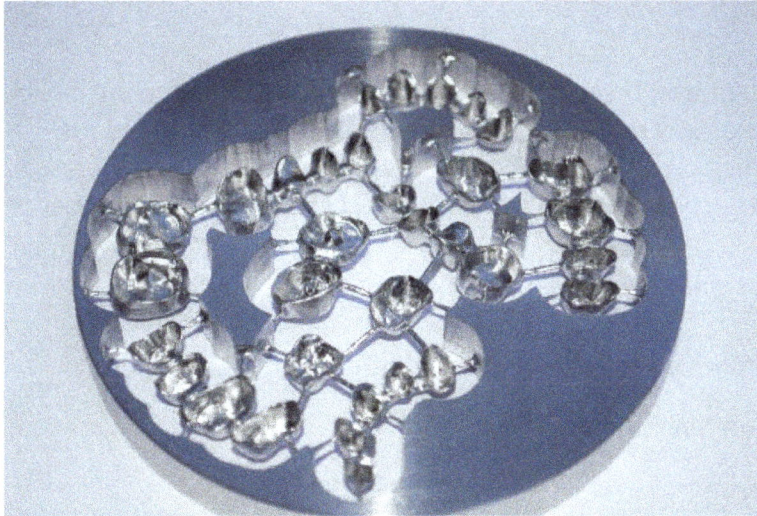

Chrome-cobalt disc with crowns for dental implants, manufactured using WorkNC CAM

History

Early commercial applications of CAM was in large companies in the automotive and aerospace industries, for example Pierre Béziers work developing the CAD/CAM application UNISURF in the 1960s for car body design and tooling at Renault.

Historically, CAM software was seen to have several shortcomings that necessitated an overly high level of involvement by skilled CNC machinists. Fallows created the first CAD software but this had severe shortcomings and was promptly taken back into the developing stage. CAM software would output code for the least capable machine, as each machine tool control added on to the standard G-code set for increased flexibility. In some cases, such as improperly set up CAM software or specific tools, the CNC machine required manual editing before the program will run properly. None of these issues were so insurmountable that a thoughtful engineer or skilled machine operator could not overcome for prototyping or small production runs; G-Code is a simple language. In high production or high precision shops, a different set of problems were encountered where an experienced CNC machinist must both hand-code programs and run CAM software.

Integration of CAD with other components of CAD/CAM/CAE Product lifecycle management (PLM) environment requires an effective CAD data exchange. Usually it had been necessary to force the CAD operator to export the data in one of the common data formats, such as IGES or STL or Parasolid formats that are supported by a wide variety of software. The output from the CAM software is usually a simple text file of G-code/M-codes, sometimes many thousands of commands long, that is then transferred to a machine tool using a direct numerical control (DNC) program or in modern Controllers using a common USB Storage Device.

CAM packages could not, and still cannot, reason as a machinist can. They could not optimize toolpaths to the extent required of mass production. Users would select the type of tool, machining process and paths to be used. While an engineer may have a working knowledge of G-code programming, small optimization and wear issues compound over time. Mass-produced items that require machining are often initially created through casting or some other non-machine method. This enables hand-written, short, and highly optimized G-code that could not be produced in a CAM package.

At least in the United States, there is a shortage of young, skilled machinists entering the workforce able to perform at the extremes of manufacturing; high precision and mass production. As CAM software and machines become more complicated, the skills required of a machinist or machine operator advance to approach that of a computer programmer and engineer rather than eliminating the CNC machinist from the workforce.

Typical areas of concern:

- High Speed Machining, including streamlining of tool paths

- Multi-function Machining

- 5 Axis Machining

- Feature recognition and machining

- Automation of Machining processes

- Ease of Use

Overcoming Historical Shortcomings

Over time, the historical shortcomings of CAM are being attenuated, both by providers of niche solutions and by providers of high-end solutions. This is occurring primarily in three arenas:

1. Ease of usage

2. Manufacturing complexity

3. Integration with PLM and the extended enterprise

Ease in Use

For the user who is just getting started as a CAM user, out-of-the-box capabilities providing Process Wizards, templates, libraries, machine tool kits, automated feature based machining and job function specific tailorable user interfaces build user confidence and speed the learning curve.

User confidence is further built on 3D visualization through a closer integration with the 3D CAD environment, including error-avoiding simulations and optimizations.

Manufacturing Complexity

The manufacturing environment is increasingly complex. The need for CAM and PLM tools by buMs are NC programmer or machinist is similar to the need for computer assistance by the pilot of modern aircraft systems. The modern machinery cannot be properly used without this assistance.

Today's CAM systems support the full range of machine tools including: turning, 5 axis machining and wire EDM. Today's CAM user can easily generate streamlined tool paths, optimized tool axis tilt for higher feed rates, better tool life and surface finish and optimized Z axis depth cuts as well as driving non-cutting operations such as the specification of probing motions.

Integration with PLM and the extended enterpriseLM to integrate manufacturing with enterprise operations from concept through field support of the finished product.

To ensure ease of use appropriate to user objectives, modern CAM solutions are scalable from a stand-alone CAM system to a fully integrated multi-CAD 3D solution-set. These solutions are created to meet the full needs of manufacturing personnel including part planning, shop documentation, resource management and data management and exchange. To prevent these solutions from detailed tool specific information a dedicated tool management

Machining Process

Most machining progresses through many stages, each of which is implemented by a variety of basic and sophisticated strategies, depending on the material and the software available.

Roughing

This process begins with raw stock, known as billet, and cuts it very roughly to shape of the final model. In milling, the result often gives the appearance of terraces, because the strategy has taken advantage of the ability to cut the model horizontally. Common strategies are zig-zag clearing, offset clearing, plunge roughing, rest-roughing.

Semi-f

This process begins with a roughed part that unevenly approximates the model and cuts to within a fixed offset distance from the model. The semi-finishing pass must leave a small amount of material so the tool can cut accurately while finishing, but not so little that the tool and material deflect instead of sending. Common strategies are raster passes, waterline passes, constant stepover passes, pencil milling.

Finishing

Finishing involves a slow pass across the material in very fine steps to produce the finished part. In finishing, the step between one pass and another is minimal. Feed rates are low and spindle speeds are raised to produce an accurate surface.

Contour milling

In milling applications on hardware with five or more axes, a separate finishing process called contouring can be performed. Instead of stepping down in fine-grained increments to approximate a surface, the work piece is rotated to make the cutting surfaces of the tool tangent to the ideal part features. This produces an excellent surface finish with high dimensional accuracy.

Software: Large Vendors

The top 20 largest CAM software companies, by direct revenues in year 2015, are sorted by global revenues:

- - CATIA, Dassault Systèmes

- - Solid Edge, Siemens

- - edgecam, work NC, surfcam Vero Software

- - HSM, Powermill, Featurecam, AUTODESK

- - CAMWorks, Geometric Technologies

- - HyperMill, OPEN MIND Technologies

- - Tebis

- - Mastercam, CNC Software

- - Cimatron, GibbsCAM 3D Systems

- - Creo, PTC

- - Vericut, CG Tech

- - TopSolid, Missler Software

- - Sprut Technology

- - FlexiSign, SAI Software

- - TYPE3, Gravotech Group

- - MecSoft Corporation

- - C&G Systems

- - SolidCAM, SolidCAM

- - NTT Data Engineering Systems

- - BobCAD-CAM

Scheduling (Production Processes)

Scheduling is the process of arranging, controlling and optimizing work and workloads in a production process or manufacturing process. Scheduling is used to allocate plant and machinery resources, plan human resources, plan production processes and purchase materials.

It is an important tool for manufacturing and engineering, where it can have a major impact on the productivity of a process. In manufacturing, the purpose of scheduling is to minimize the production time and costs, by telling a production facility when to make, with which staff, and on which equipment. Production scheduling aims to maximize the efficiency of the operation and reduce costs.

Overview

Scheduling is the process of arranging, controlling and optimizing work and workloads in a production process. Companies use backward and forward scheduling to allocate plant and machinery resources, plan human resources, plan production processes and purchase materials.

- Forward scheduling is planning the tasks from the date resources become available to determine the shipping date or the due date.

- Backward scheduling is planning the tasks from the due date or required-by date to determine the start date and/or any changes in capacity required.

The benefits of production scheduling include:

- Process change-over reduction

- Inventory reduction, leveling

- Reduced scheduling effort

- Increased production efficiency

- Labor load leveling

- Accurate delivery date quotes

- Real time information

Production scheduling tools greatly outperform older manual scheduling methods. These provide the production scheduler with powerful graphical interfaces which can be used to visually optimize real-time work loads in various stages of production, and pattern recognition allows the software to automatically create scheduling opportunities which might not be apparent without this view into the data. For example, an airline might wish to minimize the number of airport gates required for its aircraft, in order to reduce costs, and scheduling software can allow the planners to see how this can be done, by analyzing time tables, aircraft usage, or the flow of passengers.

Key Concepts in Scheduling

A key character of scheduling is the productivity, the relation between quantity of inputs and quantity of output. Key concepts here are:

- Inputs : Inputs are plant, labor, materials, tooling, energy and a clean environment.

- Outputs : Outputs are the products produced in factories either for other factories or for the end buyer. The extent to which any one product is produced within any one factory is governed by transaction cost.

- Output within the factory : The output of any one work area within the factory is an input to the next work area in that factory according to the manufacturing process. For example, the output of cutting is an input to the bending room.

- Output for the next factory : By way of example, the output of a paper mill is an input to a print factory. The output of a petrochemicals plant is an input to an asphalt plant, a cosmetics factory and a plastics factory.

- Output for the end buyer : Factory output goes to the consumer via a service business such as a retailer or an asphalt paving company.

- Resource allocation : Resource allocation is assigning inputs to produce output. The aim is to maximize output with given inputs or to minimize quantity of inputs to produce required output.

Scheduling Algorithms

Production scheduling can take a significant amount of computing power if there are a large number of tasks. Therefore, a range of short-cut algorithms (heuristics) (a.k.a. dispatching rules) are used:

- Stochastic Algorithms : Economic Lot Scheduling Problem and Economic production quantity

- Heuristic Algorithms : Modified due date scheduling heuristic and Shifting bottleneck heuristic

Batch Production Scheduling

Background

Batch production scheduling is the practice of planning and scheduling of batch manufacturing processes. Although scheduling may apply to traditionally continuous processes such as refining, it is especially important for batch processes such as those for pharmaceutical active ingredients, biotechnology processes and many specialty chemical processes. Batch production scheduling shares some concepts and techniques with finite capacity scheduling which has been applied to many manufacturing problems. The specific issues of scheduling batch manufacturing processes have generated considerable industrial and academic interest.

Scheduling in the Batch Processing Environment

A batch process can be described in terms of a recipe which comprises a bill of materials and operating instructions which describe how to make the product. The ISA S88 batch process control standard provides a framework for describing a batch process recipe. The standard provides a procedural hierarchy for a recipe. A recipe may be organized into a series of unit-procedures or major steps. Unit-procedures are organized into operations, and operations may be further organized into phases.

The following text-book recipe illustrates the organization.

- Charge and Mix materials A and B in a heated reactor, heat to 80C and react 4 hours to form C.

- Transfer to blending tank, add solvent D, Blend 1hour. Solid C precipitates.

- Centrifuge for 2 hours to separate C.

- Dry in a tray dryer for 1 hour.

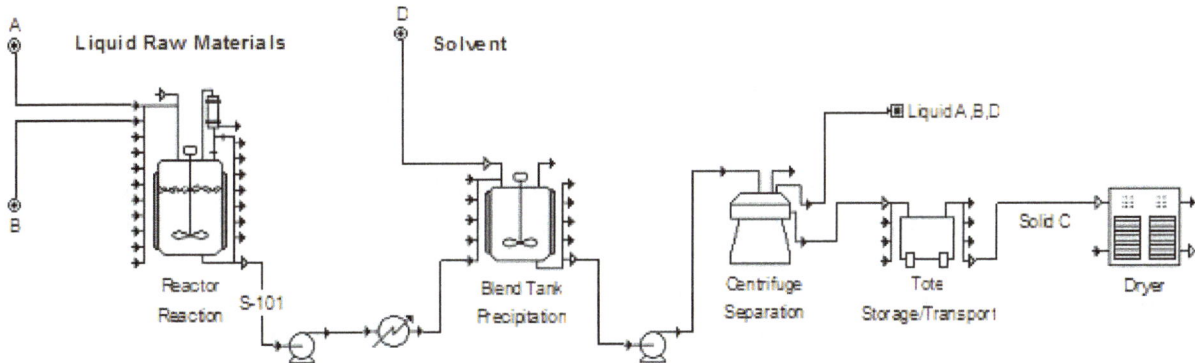

A simplified S88-style procedural organization of the recipe might appear as follows:

- Unit Procedure 1: Reaction

    o   Operation 1: Charge A & B (0.5 hours)

    o   Operation 2: Blend / Heat (1 hour)

    o   Operation 3: Hold at 80C for 4 hours

    o   Operation 4: Pump solution through cooler to blend tank (0.5 hours)

    o   Operation 5: Clean (1 hour)

- Unit Procedure 2: Blending Precipitation

    o   Operation 1: Receive solution from reactor

    o   Operation 2: Add solvent, D (0.5 hours)

    o   Operation 3: Blend for 2 hours

    o   Operation 4: Pump to centrifuge for 2 hours

    o   Operation 5: Clean up (1 hour)

- Unit Procedure 3: Centrifugation

    o   Operation 1: Centrifuge solution for 2 hours

    o   Operation 2: Clean

- Unit Procedure 4: Tote
  - Operation 1: Receive material from centrifuge
  - Operation 2: Load dryer (15 min)
- Unit Procedure 5: Dry
  - Operation 1: Load
  - Operation 2: Dry (1 hour)

Note that the organization here is intended to capture the entire process for scheduling. A recipe for process-control purposes may have a more narrow scope.

Most of the constraints and restrictions described by Pinedo are applicable in batch processing. The various operations in a recipe are subject to timing or precedence constraints that describe when they start and or end with respect to each other. Furthermore, because materials may be perishable or unstable, waiting between successive operations may be limited or impossible. Operation durations may be fixed or they may depend on the durations of other operations.

In addition to process equipment, batch process activities may require labor, materials, utilities and extra equipment.

Cycle-time Analysis

In some simple cases, an analysis of the recipe can reveal the maximum production rate and the rate limiting unit. In the process example above if a number of batches or lots of Product C are to be produced, it is useful to calculate the minimum time between consecutive batch starts (cycle-time).

The unit-procedure with the maximum duration is sometimes referred to as the bottleneck. This relationship applies when each unit-procedure has a single dedicated equipment unit.

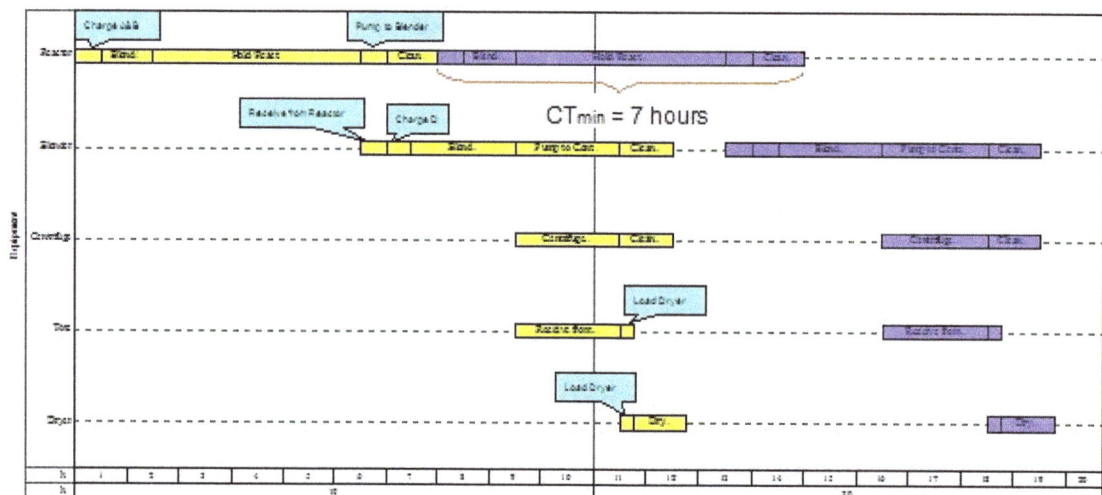

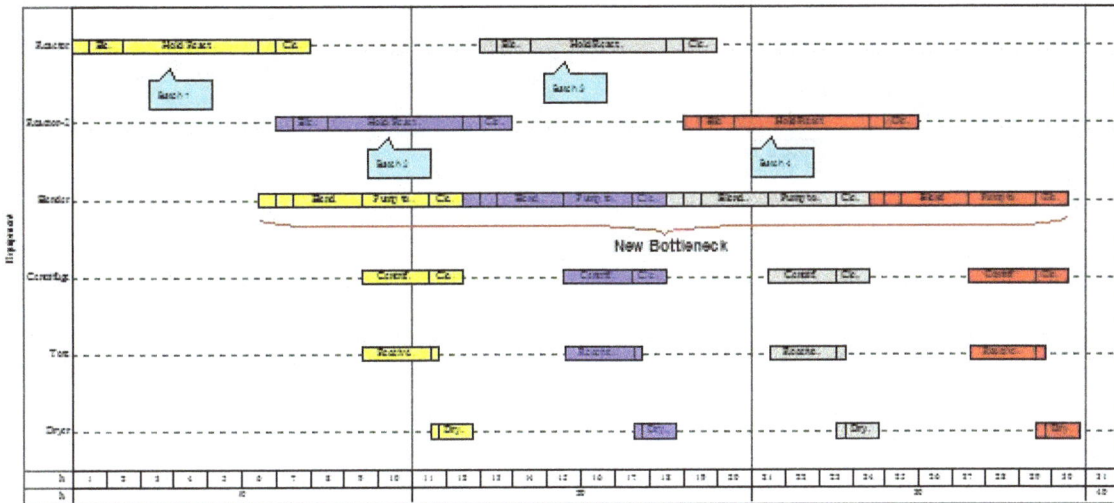

If equipment is reused within a process, the minimum cycle-time becomes more dependent on particular process details. For example, if the drying procedure in the current example is replaced with another reaction in the reactor, the minimum cycle time depends on the operating policy and on the relative durations of other procedures. In the cases below, an increase in the hold time in the tote can decrease the average minimum cycle time.

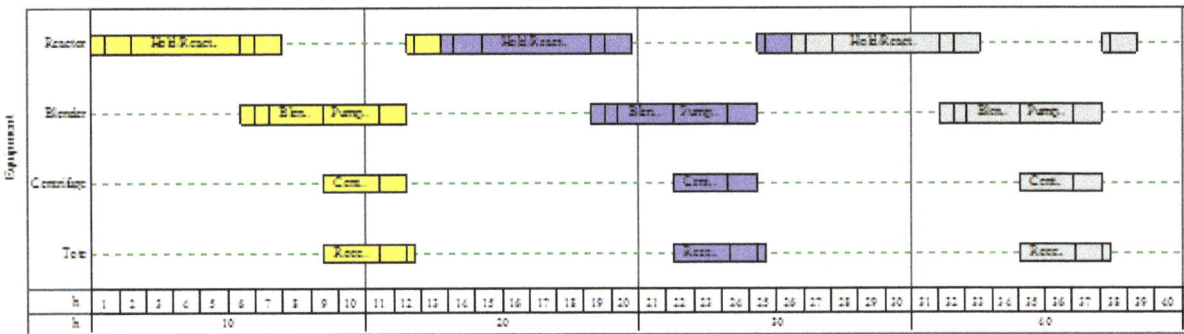

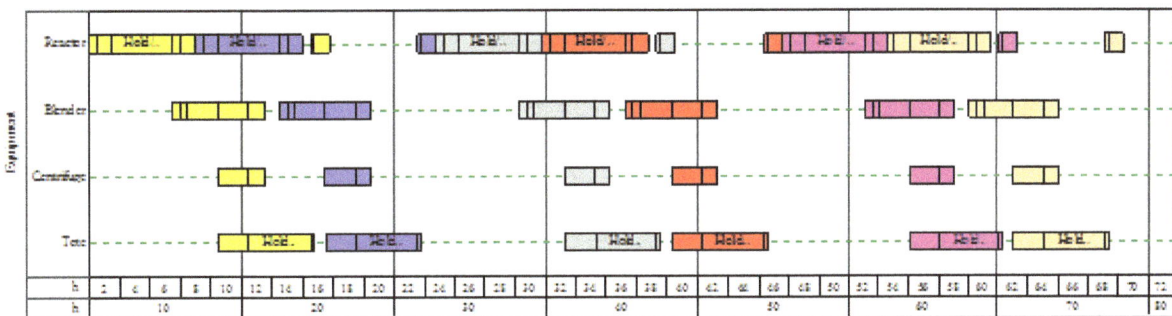

Visualization

Various charts are used to help schedulers visually manage schedules and constraints. The Gantt chart is a display that shows activities on a horizontal bar graph in which the bars represent the time of the activity. Below is an example of a Gantt chart for the process in the example described above.

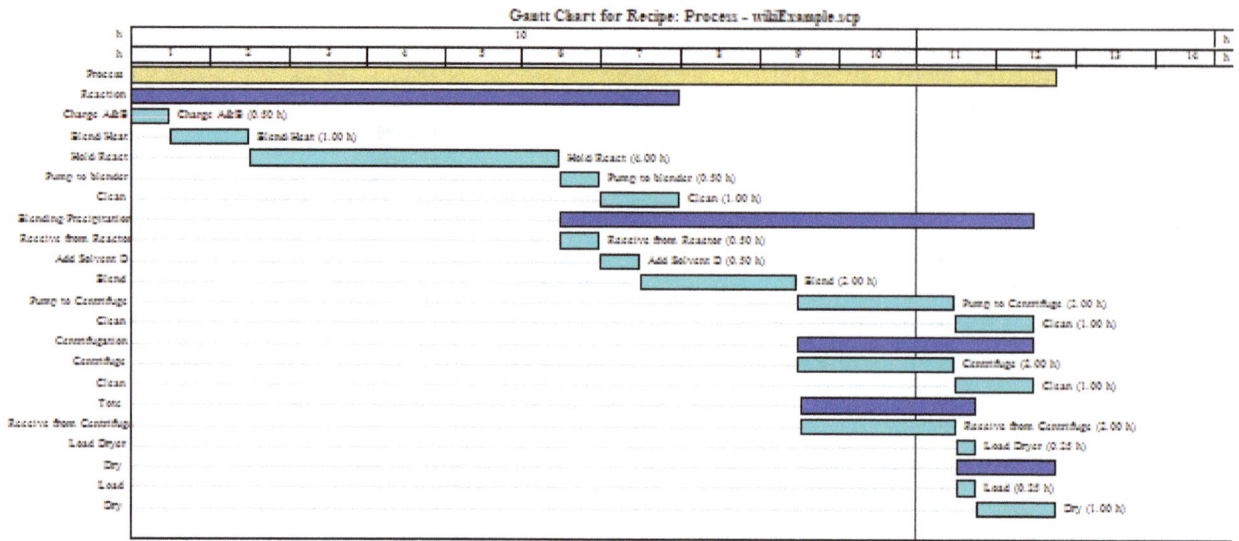

Another time chart which is also sometimes called a Gantt chart shows the time during which key resources, e.g. equipment, are occupied. The previous figures show this occupancy-style Gantt chart.

Resources that are consumed on a rate basis, e.g. electrical power, steam or labor, are generally displayed as consumption rate vs time plots.

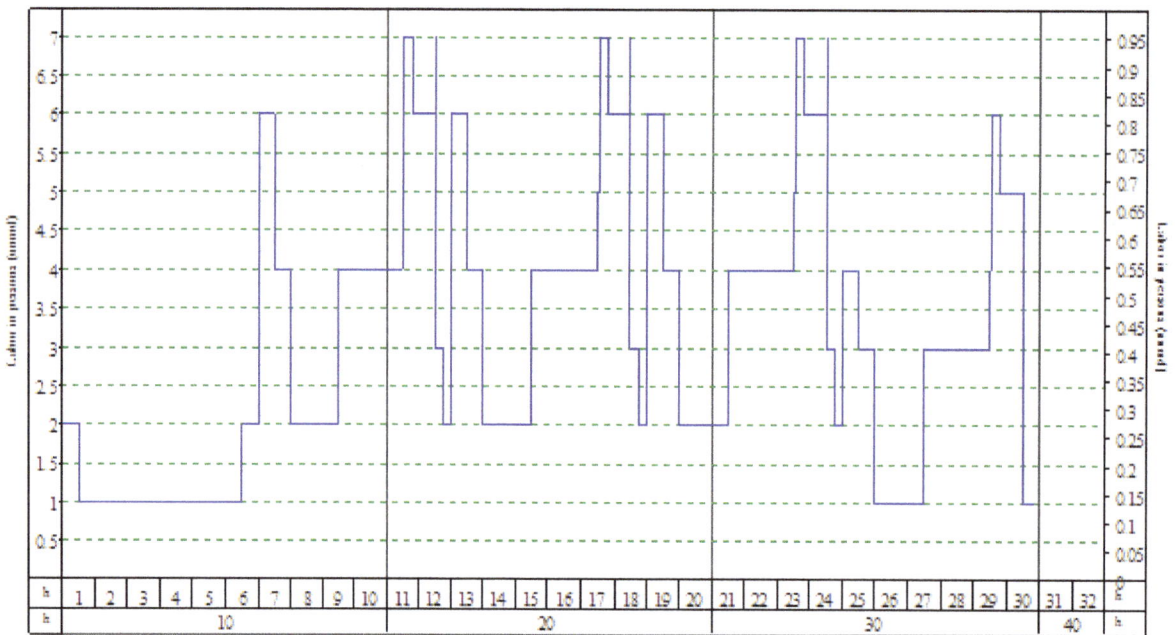

Algorithmic Methods

When scheduling situations become more complicated, for example when two or more processes share resources, it may be difficult to find the best schedule. A number of common scheduling problems, including variations on the example described above, fall into a class of problems that become very difficult to solve as their size (number of procedures and operations) grows.

A wide variety of algorithms and approaches have been applied to batch process scheduling. Early methods, which were implemented in some MRP systems assumed infinite capacity and depended only on the batch time. Such methods did not account for any resources would produce infeasible schedules.

Mathematical programming methods involve formulating the scheduling problem as an optimization problem where some objective, e.g. total duration, must be minimized (or maximized) subject to a series of constraints which are generally stated as a set of inequalities and equalities. The objective and constraints may involve zero-or-one (integer) variables as well as nonlinear relationships. An appropriate solver is applied for the resulting mixed-integer linear or nonlinear programming (MILP/MINLP) problem. The approach is theoretically guaranteed to find an optimal solution if one exists. The disadvantage is that the solver algorithm may take an unreasonable amount of time. Practitioners may use problem-specific simplifications in the formulation to get faster solutions without eliminating critical components of the scheduling model.

Constraint programming is a similar approach except that the problem is formulated only as a set of constraints and the goal is to arrive at a feasible solution rapidly. Multiple solutions are possible with this method.

References

- Freivalds, Andris; Niebel, Benjamin (2003). Methods, Standards, and Work Design (Eleventh ed.). New York: McGraw-Hill. ISBN 0072468246.

- Graham, Ben B. (2004). Detail process charting : speaking the language of process ([Online-Ausg.] ed.). Hoboken, N.J.: Wiley. p. 2. ISBN 9780471653943.

- Groover, M. P. Work Systems and Methods, measurement, and Management of Work. Pearson Education Inter-national, 2007 ISBN 978-0-13-140650-6

- Miller, Doug, Towards Sustainable Labour Costing in UK Fashion Retail (February 5, 2013). Available at SSRN: doi:10.2139/ssrn.2212100

- Robinson, M. A. (2010). "Work sampling: Methodological advances and new applications". Human Factors and Ergonomics in Manufacturing & Service Industries. 20 (1): 42–60. doi:10.1002/hfm.20186.

Industrial Engineering: Research and Management

Research and management are significant areas of study under the branch of industrial engineering. The following chapter unfolds its crucial aspects in a critical yet systematic manner. Major elements like operations research and operations management are discussed in this chapter.

Operations Research

Operations research, or operational research in British usage, is a discipline that deals with the application of advanced analytical methods to help make better decisions. Further, the term 'operational analysis' is used in the British (and some British Commonwealth) military, as an intrinsic part of capability development, management and assurance. In particular, operational analysis forms part of the Combined Operational Effectiveness and Investment Appraisals (COEIA), which support British defence capability acquisition decision-making.

It is often considered to be a sub-field of mathematics. The terms management science and decision science are sometimes used as synonyms.

Employing techniques from other mathematical sciences, such as mathematical modeling, statistical analysis, and mathematical optimization, operations research arrives at optimal or near-optimal solutions to complex decision-making problems. Because of its emphasis on human-technology interaction and because of its focus on practical applications, operations research has overlap with other disciplines, notably industrial engineering and operations management, and draws on psychology and organization science. Operations research is often concerned with determining the maximum (of profit, performance, or yield) or minimum (of loss, risk, or cost) of some real-world objective. Originating in military efforts before World War II, its techniques have grown to concern problems in a variety of industries.

Overview

Operational research (OR) encompasses a wide range of problem-solving techniques and methods applied in the pursuit of improved decision-making and efficiency, such as simulation, mathematical optimization, queueing theory and other stochastic-process models, Markov decision processes, econometric methods, data envelopment analysis, neural networks, expert systems, decision analysis, and the analytic hierarchy process. Nearly all of these techniques involve the construction of mathematical models that attempt to describe the system. Because of the computational and statistical nature of most of these fields, OR also has strong ties to computer science and analytics. Operational researchers faced with a new problem must determine which of these techniques are most appropriate given the nature of the system, the goals for improvement, and constraints on time and computing power.

The major subdisciplines in modern operational research, as identified by the journal *Operations Research*, are:

- Computing and information technologies
- Financial engineering
- Manufacturing, service sciences, and supply chain management
- Policy modeling and public sector work
- Revenue management
- Simulation
- Stochastic models
- Transportation

History

As a discipline, operational research originated in the efforts of military planners during World War I (convoy theory and Lanchester's laws). In the decades after the two world wars, the techniques were more widely applied to problems in business, industry and society. Since that time, operational research has expanded into a field widely used in industries ranging from petrochemicals to airlines, finance, logistics, and government, moving to a focus on the development of mathematical models that can be used to analyse and optimize complex systems, and has become an area of active academic and industrial research.

Historical Origins

Early work in operational research was carried out by individuals such as Charles Babbage. His research into the cost of transportation and sorting of mail led to England's universal "Penny Post" in 1840, and studies into the dynamical behaviour of railway vehicles in defence of the GWR's broad gauge. Percy Bridgman brought operational research to bear on problems in physics in the 1920s and would later attempt to extend these to the social sciences.

Modern operational research originated at the Bawdsey Research Station in the UK in 1937 and was the result of an initiative of the station's superintendent, A. P. Rowe. Rowe conceived the idea as a means to analyse and improve the working of the UK's early warning radar system, Chain Home (CH). Initially, he analysed the operating of the radar equipment and its communication networks, expanding later to include the operating personnel's behaviour. This revealed unappreciated limitations of the CH network and allowed remedial action to be taken.

Scientists in the United Kingdom including Patrick Blackett (later Lord Blackett OM PRS), Cecil Gordon, Solly Zuckerman, (later Baron Zuckerman OM, KCB, FRS), C. H. Waddington, Owen Wansbrough-Jones, Frank Yates, Jacob Bronowski and Freeman Dyson, and in the United States with George Dantzig looked for ways to make better decisions in such areas as logistics and training schedules

Second World War

The modern field of operational research arose during World War II. In the World War II era, operational research was defined as "a scientific method of providing executive departments with a quantitative basis for decisions regarding the operations under their control." Other names for it included operational analysis (UK Ministry of Defence from 1962) and quantitative management.

During the Second World War close to 1,000 men and women in Britain were engaged in operational research. About 200 operational research scientists worked for the British Army.

Patrick Blackett worked for several different organizations during the war. Early in the war while working for the Royal Aircraft Establishment (RAE) he set up a team known as the "Circus" which helped to reduce the number of anti-aircraft artillery rounds needed to shoot down an enemy aircraft from an average of over 20,000 at the start of the Battle of Britain to 4,000 in 1941.

In 1941 Blackett moved from the RAE to the Navy, after first working with RAF Coastal Command, in 1941 and then early in 1942 to the Admiralty. Blackett's team at Coastal Command's Operational Research Section (CC-ORS) included two future Nobel prize winners and many other people who went on to be pre-eminent in their fields. They undertook a number of crucial analyses that aided the war effort. Britain introduced the convoy system to reduce shipping losses, but while the principle of using warships to accompany merchant ships was generally accepted, it was unclear whether it was better for convoys to be small or large. Convoys travel at the speed of the slowest member, so small convoys can travel faster. It was also argued that small convoys would be harder for German U-boats to detect. On the other hand, large convoys could deploy more warships against an attacker. Blackett's staff showed that the losses suffered by convoys depended largely on the number of escort vessels present, rather than the size of the convoy. Their conclusion was that a few large convoys are more defensible than many small ones.

A Liberator in standard RAF green/dark earth/black night bomber finish as originally used by Coastal Command

While performing an analysis of the methods used by RAF Coastal Command to hunt and destroy submarines, one of the analysts asked what colour the aircraft were. As most of them were from

Bomber Command they were painted black for night-time operations. At the suggestion of CC-ORS a test was run to see if that was the best colour to camouflage the aircraft for daytime operations in the grey North Atlantic skies. Tests showed that aircraft painted white were on average not spotted until they were 20% closer than those painted black. This change indicated that 30% more submarines would be attacked and sunk for the same number of sightings. As a result of these findings Coastal Command changed their aircraft to using white undersurfaces.

A Warwick in the revised RAF Coastal Command green/dark grey/white finish

Other work by the CC-ORS indicated that on average if the trigger depth of aerial-delivered depth charges (DCs) were changed from 100 feet to 25 feet, the kill ratios would go up. The reason was that if a U-boat saw an aircraft only shortly before it arrived over the target then at 100 feet the charges would do no damage (because the U-boat wouldn't have had time to descend as far as 100 feet), and if it saw the aircraft a long way from the target it had time to alter course under water so the chances of it being within the 20-foot kill zone of the charges was small. It was more efficient to attack those submarines close to the surface when the targets' locations were better known than to attempt their destruction at greater depths when their positions could only be guessed. Before the change of settings from 100 feet to 25 feet, 1% of submerged U-boats were sunk and 14% damaged. After the change, 7% were sunk and 11% damaged. (If submarines were caught on the surface, even if attacked shortly after submerging, the numbers rose to 11% sunk and 15% damaged). Blackett observed "there can be few cases where such a great operational gain had been obtained by such a small and simple change of tactics".

Bomber Command's Operational Research Section (BC-ORS), analysed a report of a survey carried out by RAF Bomber Command. For the survey, Bomber Command inspected all bombers returning from bombing raids over Germany over a particular period. All damage inflicted by German air defences was noted and the recommendation was given that armour be added in the most heavily damaged areas. This recommendation was not adopted because the fact that the aircraft returned with these areas damaged indicated these areas were not vital, and adding armour to non-vital areas where damage is acceptable negatively affects aircraft performance. Their suggestion to remove some of the crew so that an aircraft loss would result in fewer personnel losses, was also rejected by RAF command. Blackett's team made the logical recommendation that the armour be placed in the areas which were completely untouched by damage in the bombers which returned. They reasoned that the survey was biased, since it only included aircraft that returned to Britain. The untouched areas of returning aircraft were probably vital areas, which, if hit, would result in the loss of the aircraft.

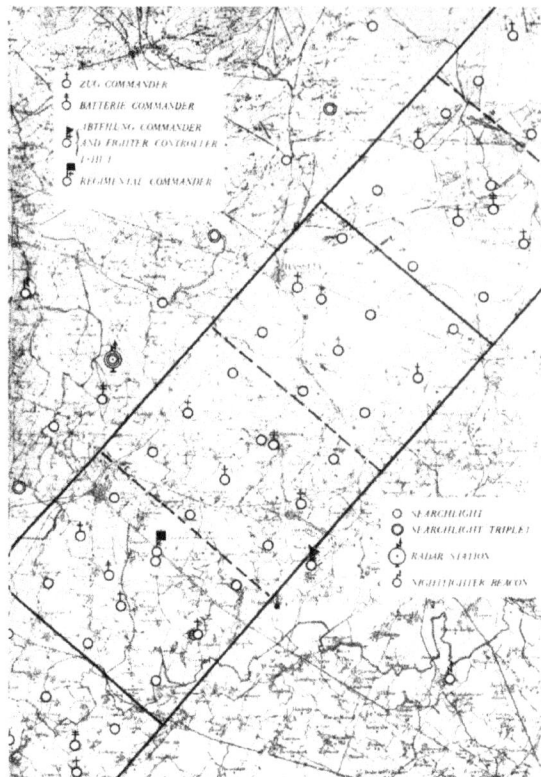

Map of *Kammhuber Line*

When Germany organised its air defences into the Kammhuber Line, it was realised by the British that if the RAF bombers were to fly in a bomber stream they could overwhelm the night fighters who flew in individual cells directed to their targets by ground controllers. It was then a matter of calculating the statistical loss from collisions against the statistical loss from night fighters to calculate how close the bombers should fly to minimise RAF losses.

The "exchange rate" ratio of output to input was a characteristic feature of operational research. By comparing the number of flying hours put in by Allied aircraft to the number of U-boat sightings in a given area, it was possible to redistribute aircraft to more productive patrol areas. Comparison of exchange rates established "effectiveness ratios" useful in planning. The ratio of 60 mines laid per ship sunk was common to several campaigns: German mines in British ports, British mines on German routes, and United States mines in Japanese routes.

Operational research doubled the on-target bomb rate of B-29s bombing Japan from the Marianas Islands by increasing the training ratio from 4 to 10 percent of flying hours; revealed that wolfpacks of three United States submarines were the most effective number to enable all members of the pack to engage targets discovered on their individual patrol stations; revealed that glossy enamel paint was more effective camouflage for night fighters than traditional dull camouflage paint finish, and the smooth paint finish increased airspeed by reducing skin friction.

On land, the operational research sections of the Army Operational Research Group (AORG) of the Ministry of Supply (MoS) were landed in Normandy in 1944, and they followed British forces in the advance across Europe. They analysed, among other topics, the effectiveness of artillery, aerial bombing and anti-tank shooting.

After World War II

With expanded techniques and growing awareness of the field at the close of the war, operational research was no longer limited to only operational, but was extended to encompass equipment procurement, training, logistics and infrastructure. Operations Research also grew in many areas other than the military once scientists learned to apply its principles to the civilian sector. With the development of the simplex algorithm for linear programming in 1947 and the development of computers over the next three decades, Operations Research can now "solve problems with hundreds of thousands of variables and constraints. Moreover, the large volumes of data required for such problems can be stored and manipulated very efficiently."

Problems Addressed

- Critical path analysis or project planning: identifying those processes in a complex project which affect the overall duration of the project

- Floorplanning: designing the layout of equipment in a factory or components on a computer chip to reduce manufacturing time (therefore reducing cost)

- Network optimization: for instance, setup of telecommunications networks to maintain quality of service during outages

- Allocation problems

- Facility location

- Assignment Problems:

    o Assignment problem

    o Generalized assignment problem

    o Quadratic assignment problem

    o Weapon target assignment problem

- Bayesian search theory: looking for a target

- Optimal search

- Routing, such as determining the routes of buses so that as few buses are needed as possible

- Supply chain management: managing the flow of raw materials and products based on uncertain demand for the finished products

- Efficient messaging and customer response tactics

- Automation: automating or integrating robotic systems in human-driven operations processes

- Globalization: globalizing operations processes in order to take advantage of cheaper materials, labor, land or other productivity inputs

- Transportation: managing freight transportation and delivery systems (Examples: LTL shipping, intermodal freight transport, travelling salesman problem)

- Scheduling:

  o Personnel staffing

  o Manufacturing steps

  o Project tasks

  o Network data traffic: these are known as queueing models or queueing systems.

  o Sports events and their television coverage

- Blending of raw materials in oil refineries

- Determining optimal prices, in many retail and B2B settings, within the disciplines of pricing science

Operational research is also used extensively in government where evidence-based policy is used.

Management Science

In 1967 Stafford Beer characterized the field of management science as "the business use of operations research". However, in modern times the term management science may also be used to refer to the separate fields of organizational studies or corporate strategy. Like operational research itself, management science (MS) is an interdisciplinary branch of applied mathematics devoted to optimal decision planning, with strong links with economics, business, engineering, and other sciences. It uses various scientific research-based principles, strategies, and analytical methods including mathematical modeling, statistics and numerical algorithms to improve an organization's ability to enact rational and meaningful management decisions by arriving at optimal or near optimal solutions to complex decision problems. In short, management sciences help businesses to achieve their goals using the scientific methods of operational research.

The management scientist's mandate is to use rational, systematic, science-based techniques to inform and improve decisions of all kinds. Of course, the techniques of management science are not restricted to business applications but may be applied to military, medical, public administration, charitable groups, political groups or community groups.

Management science is concerned with developing and applying models and concepts that may prove useful in helping to illuminate management issues and solve managerial problems, as well as designing and developing new and better models of organizational excellence.

The application of these models within the corporate sector became known as management science.

Related Fields

Some of the fields that have considerable overlap with Operations Research and Management Science include:

• Business Analytics	• Logistics
• Data mining	• Mathematical modeling
• Decision analysis	• Mathematical optimization
• Engineering	• Probability and statistics
• Financial engineering	• Project management
• Forecasting	• Policy analysis
• Game theory	• Simulation
• Graph theory	• Social network/Transportation forecasting models
• Industrial engineering	• Stochastic processes
	• Supply chain management

Applications

Applications of management science are abundant such as in airlines, manufacturing companies, service organizations, military branches, and government. The range of problems and issues to which management science has contributed insights and solutions is vast. It includes:

- Scheduling airlines, trains, buses etc.

- Assignment (assigning crew to flights, trains or buses; employees to projects)

- Facility location (deciding the most appropriate location for the new facilities such as a warehouse, factory or fire station)

- Network flows (managing the flow of water from reservoirs)

- Health service (information and supply chain management for health services)

- Game theory (identifying, understanding and developing the strategies adopted by companies)

Management science is also concerned with so-called "soft-operational analysis", which concerns methods for strategic planning, strategic decision support, and problem structuring methods. In dealing with these sorts of challenges mathematical modeling and simulation are not appropriate or will not suffice. Therefore, during the past 30 years, a number of non-quantified modeling methods have been developed. These include:

- stakeholder based approaches including metagame analysis and drama theory

- morphological analysis and various forms of influence diagrams

- approaches using cognitive mapping

- the strategic choice approach

- robustness analysis

Societies

The International Federation of Operational Research Societies (IFORS) is an umbrella organization for operational research societies worldwide, representing approximately 50 national societies including those in the US, UK, France, Germany, Canada, Australia, New Zealand, Philippines, India, Japan and South Africa. The constituent members of IFORS form regional groups, such as that in Europe. Other important operational research organizations are Simulation Interoperability Standards Organization (SISO) and Interservice/Industry Training, Simulation and Education Conference (I/ITSEC)

In 2004 the US-based organization INFORMS began an initiative to market the OR profession better, including a website entitled *The Science of Better* which provides an introduction to OR and examples of successful applications of OR to industrial problems. This initiative has been adopted by the Operational Research Society in the UK, including a website entitled *Learn about OR*.

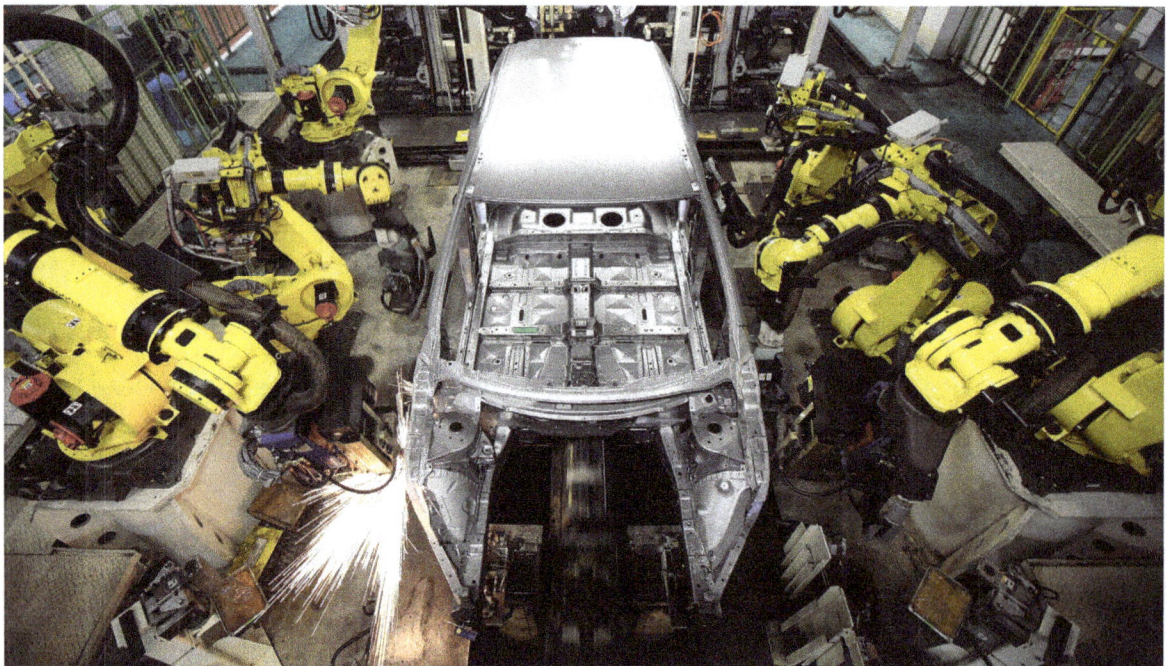

Operations Management

Operations management is an area of management concerned with designing and controlling the process of production and redesigning business operations in the production of goods or services. It involves the responsibility of ensuring that business operations are efficient in terms of using as few resources as needed and effective in terms of meeting customer requirements. It is concerned with managing the process that converts inputs (in the forms of raw materials, labor, and energy) into outputs (in the form of goods and/or services). The relationship of operations management to senior management in commercial contexts can be compared to the relationship of line officers to highest-level senior officers in military science. The highest-level officers shape the strategy and revise it over time, while the line officers make tactical decisions in support of carrying out the strategy. In business as in military affairs, the boundaries between levels are not always distinct; tactical information dynamically informs strategy, and individual people often move between roles over time.

Ford Motor car assembly line: the classical example of a manufacturing production system. Post office queue. Operations management studies both manufacturing and services.

According to the United States Department of Education, operations management is the field concerned with managing and directing the physical and/or technical functions of a firm or organization, particularly those relating to development, production, and manufacturing. Operations management programs typically include instruction in principles of general management, manufacturing and production systems, factory management, equipment maintenance management, production control, industrial labor relations and skilled trades supervision, strategic manufacturing policy, systems analysis, productivity analysis and cost control, and materials planning.

History

The history of production and operation systems began around 5000 B.C. when Sumerian priests developed the ancient system of recording inventories, loans, taxes, and business transactions. The next major historical application of operation systems occurred in 4000 B.C. It was during this time that the Egyptians started using planning, organization, and control in large projects such as the construction of the pyramids. By 1100 B.C., labor was being specialized in China; by about 370 B.C., Xenophon described the advantages of dividing the various operations necessary for the production of shoes among different individuals in ancient Greece.

Shoemakers, 1568

In the Middle Ages, kings and queens ruled over large areas of land. Loyal noblemen maintained large sections of the monarch's territory. This hierarchical organization in which people were divided into classes based on social position and wealth became known as the feudal system. In the feudal system, vassals and serfs produced for themselves and people of higher classes by using the ruler's land and resources. Although a large part of labor was employed in agriculture, artisans contributed to economic output and formed guilds. The guild system, operating mainly between 1100 and 1500, consisted of two types: merchant guilds, who bought and sold goods, and craft guilds, which made goods. Although guilds were regulated as to the quality of work performed, the resulting system was rather rigid, shoemakers, for example, were prohibited from tannin hides.

The industrial revolution was facilitated by two elements: interchangeability of parts and division of labor. Division of labor has always been a feature from the beginning of civilization, the extent to which the division is carried out varied considerably depending on period and location. Compared to the Middle Ages, the Renaissance and the Age of Discovery were characterized by a greater specialization in labor, one of the characteristics of growing European cities and trade. It was in the late eighteenth century that Eli Whitney popularized the concept of interchangeability of parts when he manufactured 10,000 muskets. Up to this point in the history of manufacturing, each product (e.g. each gun) was considered a special order, meaning that parts of a given gun were fitted only for that particular gun and could not be used in other guns. Interchangeability of parts allowed the mass production of parts independent of the final products in which they would be used.

In 1883, Frederick Winslow Taylor introduced the stopwatch method for accurately measuring the time to perform each single task of a complicated job. He developed the scientific study of productivity and identifying how to coordinate different tasks to eliminate wasting of time and increase the quality of work. The next generation of scientific study occurred with the development of work sampling and predetermined motion time systems (PMTS). Work sampling is used to measure the random variable associated with the time of each task. PMTS allows the use of standard predetermined tables of the smallest body movements (e.g. turning the left wrist by 90°), and integrating them to predict the time needed to perform a simple task. PMTS has gained substantial importance due to the fact that it can predict work measurements without observing the actual work. The foundation of PMTS was laid out by the research and development of Frank B. and Lillian M. Gilbreth around 1912. The Gilbreths took advantage of taking motion pictures at known time intervals while operators were performing the given task.

The idea of the production line has been used multiple times in history prior to Henry Ford: the Venetian Arsenal (1104); Smith's pin manufacturing, in the Wealth of Nations (1776) or Brunel's Portsmouth Block Mills (1802). Ransom Olds was the first to manufacture cars using the assembly line system, but Henry Ford developed the first auto assembly system where a car chassis was moved through the assembly line by a conveyor belt while workers added components to it until the car was completed. During World War II, the growth of computing power led to further development of efficient manufacturing methods and the use of advanced mathematical and statistical tools. This was supported by the development of academic programs in industrial and systems engineering disciplines, as well as fields of operations research and management science (as multi-disciplinary fields of problem solving). While systems engineering concentrated on the broad characteristics of the relationships between inputs and outputs of generic systems, operations researchers concentrated on solving specific and focused problems. The synergy of operations research and systems engineering allowed for the realization of solving large scale and complex problems in the modern era. Recently, the development of faster and smaller computers, intelligent systems, and the World Wide Web has opened new opportunities for operations, manufacturing, production, and service systems.

Malakooti (2013) states that production and operation systems can be divided into five phases:

1. Empiricism (learning from experience)

2. Analysis (scientific management)

3. Synthesis (development of mathematical problem solving tools)

4. Isolated Systems with Single Objective (use of Integrated and Intelligent Systems, and WWW)

5. Integrated Complex Systems with Multiple Objectives (development of ecologically sound systems, environmentally sustainable systems, considering individual preferences)

Industrial Revolution

Marshall's flax mill in Holbeck. The textile industry is the prototypical example of the English industrial revolution.

Before the First industrial revolution work was mainly done through two systems: domestic system and craft guilds. In the domestic system merchants took materials to homes where artisans performed the necessary work, craft guilds on the other hand were associations of artisans which passed work from one shop to another, for example: leather was tanned by a tanner, passed to curriers, and finally arrived at shoemakers and saddlers.

The beginning of the industrial revolution is usually associated with 18th century English textile industry, with the invention of flying shuttle by John Kay in 1733, the spinning jenny by James Hargreaves in 1765, the water frame by Richard Arkwright in 1769 and the steam engine by James Watt in 1765. In 1851 at the Crystal Palace Exhibition the term American system of manufacturing was used to describe the new approach that was evolving in the United States of America which was based on two central features: interchangeable parts and extensive use of mechanization to produce them.

Henry Ford was 39 years old when he founded the Ford Motor Company in 1903, with $28,000 capital from twelve investors. The model T car was introduced in 1908, however it was not until Ford implemented the assembly line concept, that his vision of making a popular car affordable by every middle-class American citizen would be realized. The first factory in which Henry Ford used the concept of the assembly line was Highland Park (1913), he characterized the system as follows:

"The thing is to keep everything in motion and take the work to the man and not the man to the work. That is the real principle of our production, and conveyors are only one of many means to an end"

This became one the central ideas that led to mass production, one of the main elements of the Second Industrial Revolution, along with emergence of the electrical industry and petroleum industry.

Operations Management

Although productivity benefited considerably from technological inventions and division of labour, the problem of systematic measurement of performances and the calculation of these by the use of formulas remained somewhat unexplored until Frederick Taylor, whose early work focused on developing what he called a "differential piece-rate system" and a series of experiments, measurements and formulas dealing with cutting metals and manual labor. The differential piece-rate system consisted in offering two different pay rates for doing a job: a higher rate for workers with high productivity (efficiency) and who produced high quality goods (effectiveness) and a lower rate for those who fail to achieve the standard. One of the problems Taylor believed could be solved with this system, was the problem of soldiering: faster workers reducing their production rate to that of the slowest worker. In 1911 Taylor published his "The Principles of Scientific Management", in which he characterized scientific management (also known as Taylorism) as:

1. The development of a true science;

2. The scientific selection of the worker;

3. The scientific education and development of the worker;

4. Intimate friendly cooperation between the management and the workers.

Taylor is also credited for developing stopwatch time study, this combined with Frank and Lillian Gilbreth motion study gave way to time and motion study which is centered on the concepts of standard method and standard time. Frank Gilbreth is also responsible for introducing the flow process chart in 1921. Other contemporaries of Taylor worth remembering are Morris Cooke (rural electrification in the 1920s and implementer of Taylor's principles of scientific management in the Philadelphia's Department of Public Works), Carl Barth (speed-and-feed-calculating slide rules) and Henry Gantt (Gantt chart). Also in 1910 Hugo Diemer published the first industrial engineering book: Factory Organization and Administration.

In 1913 Ford Whitman Harris published his "How Many parts to make at once" in which he presented the idea of the economic order quantity model. He described the problem as follows:

"Interest on capital tied up in wages, material and overhead sets a maximum limit to the quantity of parts which can be profitably manufactured at one time; "setup costs" on the job fix the minimum. Experience has shown one manager a way to determine the economical size of lots"

This paper inspired a large body of mathematical literature focusing on the problem of production planning and inventory control.

In 1924 Walter Shewhart introduced the control chart through a technical memorandum while working at Bell Labs, central to his method was the distinction between common cause and special cause of variation. In 1931 Shewhart published his Economic Control of Quality of Manufactured Product, the first systematic treatment of the subject of Statistical Process Control (SPC).

In the 1940s methods-time measurement (MTM) was developed by H.B. Maynard, JL Schwab and GJ Stegemerten. MTM was the first of a series of predetermined motion time systems, predetermined in the sense that estimates of time are not determined in loco but are derived from an industry standard. This was explained by its originators in a book they published in 1948 called "Method-Time Measurement".

Up to this point in history, optimization techniques were known for a very long time, from the simple methods employed by F.W.Harris to the more elaborate techniques of the calculus of variations developed by Euler in 1733 or the multipliers employed by Lagrange in 1811, and computers were slowly being developed, first as analog computers by Sir William Thomson (1872) and James Thomson (1876) moving to the eletromechanical computers of Konrad Zuse (1939 and 1941). During World War II however, the development of mathematical optimization went through a major boost with the development of the Colossus computer, the first electronic digital computer that was all programmable, and the possibility to computationally solve large linear programming problems, first by Kantorovich in 1939 working for the Soviet government and latter on in 1947 with the simplex method of Dantzig. These methods are known today as belonging to the field of operations research.

From this point on a curious development took place: while in the United States the possibility of applying the computer to business operations led to the development of management software architecture such as MRP and successive modifications, and ever more sophisticated optimization techniques and manufacturing simulation software, in post-war Japan a series of events at Toyota Motor led to the development of the Toyota Production System (TPS) and Lean Manufacturing.

In 1943, in Japan, Taiichi Ohno arrived at Toyota Motor company. Toyota evolved a unique manufacturing system centered on two complementary notions: just in time (produce only what is needed) and autonomation (automation with a human touch). Regarding JIT, Ohno was inspired by American supermarkets: workstations functioned like a supermarket shelf where the customer can get products they need, at the time they need and in the amount needed, the workstation (shelf) is then restocked. Autonomation was developed by Toyoda Sakichi in Toyoda Spinning and Weaving: an automatically activated loom that was also foolproof, that is automatically detected problems. In 1983 J.N Edwards published his "MRP and Kanban-American style" in which he described JIT goals in terms of seven zeros: zero defects, zero (excess) lot size, zero setups, zero breakdowns, zero handling, zero lead time and zero surging. This period also marks the spread of Total Quality Management (TQM) in Japan, ideas initially developed by American authors such as Deming, Juran and Armand V. Feigenbaum. TQM is a strategy for implementing and managing quality improvement on an organizational basis, this includes: participation, work culture, customer focus, supplier quality improvement and integration of the quality system with business goals. Schnonberger identified seven fundamentals principles essential to the Japanese approach:

1. Process control: SPC and worker responsibility over quality

2. Easy able -to-see quality: boards, gauges, meters, etc. and poka-yoke

3. Insistence on compliance: "quality first"

4. Line stop: stop the line to correct quality problems

5. Correcting one's own errors: worker fixed a defective part if he produced it

6. The 100% check: automated inspection techniques and foolproof machines

7. Continual improvement: ideally zero defects

In 1987 the International Organization for Standardization (ISO), recognizing the growing importance of quality, issued the ISO 9000, a family of standards related to quality management systems. There has been some controversy though regarding the proper procedures to follow and the amount of paperwork involved.

Meanwhile, in the sixties, a different approach was developed by George W. Plossl and Oliver W. Wight, this approach was continued by Joseph Orlicky as a response to the TOYOTA Manufacturing Program which led to Material Requirements Planning (MRP) at IBM, latter gaining momentum in 1972 when the American Production and Inventory Control Society launched the "MRP Crusade". One of the key insights of this management system was the distinction between dependent demand and independent demand. Independent demand is demand which originates outside of the production system, therefore not directly controllable, and dependent demand is demand for components of final products, therefore subject to being directly controllable by management through the bill of materials, via product design. Orlicky wrote "Materials Requirement Planning" in 1975, the first hard cover book on the subject. MRP II was developed by Gene Thomas at IBM, and expanded the original MRP software to include additional production functions. Enterprise resource planning (ERP) is the modern software architecture, which addresses, besides production operations, distribution, accounting, human resources and procurement.

Recent trends in the field revolve around concepts such as:

- Business Process Re-engineering (launched by Michael Hammer in 1993): a business management strategy focusing on the analysis and design of workflows and business processes within an organization. BPR seeks to help companies radically restructure their organizations by focusing on the ground-up design of their business processes.

- Lean Manufacturing: a systemic method for the elimination of waste ("Muda") within a manufacturing process. Lean also takes into account waste created through overburden ("Muri") and waste created through unevenness in work loads ("Mura"). The term lean manufacturing was coined in the book The Machine that Changed the World.

- Six Sigma (an approach to quality developed at Motorola between 1985-1987): Six Sigma refers to control limits placed at six (6) standard deviations from the mean of a normal distribution, this became very famous after Jack Welch of General Electric launched a company-wide initiative in 1995 to adopt this set of methods. More recently, Six Sigma has included DMAIC (for improving processes) and DFSS (for designing new products and new processes)

- Reconfigurable Manufacturing Systems: a production system designed at the outset for rapid change in its structure, as well as its hardware and software components, in order to quickly adjust its production capacity and functionality within a part family in response to sudden market changes or intrinsic system change.

Topics

Production Systems

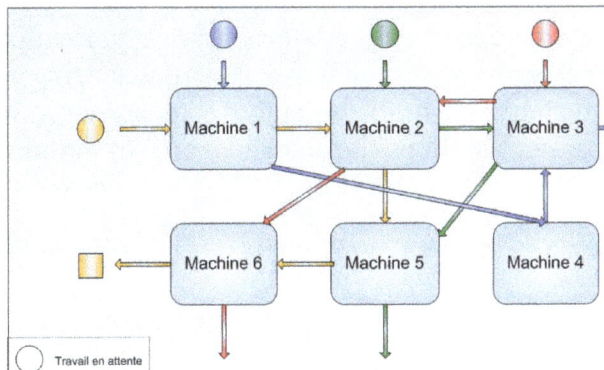

In a job shop machines are grouped by technological similarities regarding transformation processes, therefore a single shop can work very different products (in this picture four colors). Also notice that in this drawing each shop contains a single machine.

Flexible Manufacturing System: in the middle there are two rails for the shuttle to move pallets between machining centers (there are also FMS which use AGVs), in front of each machining center there is a buffer and in left we have a shelf for storing pallets. Usually in the back there is a similar system for managing the set of tools required for different machining operations.

A production system comprises both the technological elements (machines and tools) and organizational behavior (division of labor and information flow). An individual production system is usually analyzed in the literature referring to a single business, therefore it's usually improper to include in a given production system the operations necessary to process goods that are obtained by purchasing or the operations carried by the customer on the sold products, the reason being simply that since businesses need to design their own production systems this then becomes the focus of analysis, modeling and decision making (also called "configuring" a production system) .

A first possible distinction in production systems (technological classification) is between continuous process production and discrete part production (manufacturing).

- Process production means that the product undergoes physical-chemical transformations and lacks assembly operations, therefore the original raw materials can't easily be obtained from the final product, examples include: paper, cement, nylon and petroleum products.

- Part production (ex:cars and ovens) comprises both fabrication systems and assembly systems. In the first category we find job shops, manufacturing cells, flexible manufacturing systems and transfer lines, in the assembly category we have fixed position systems, assembly lines and assembly shops (both manual and/or automated operations).

Classificazione di *Wortmann*

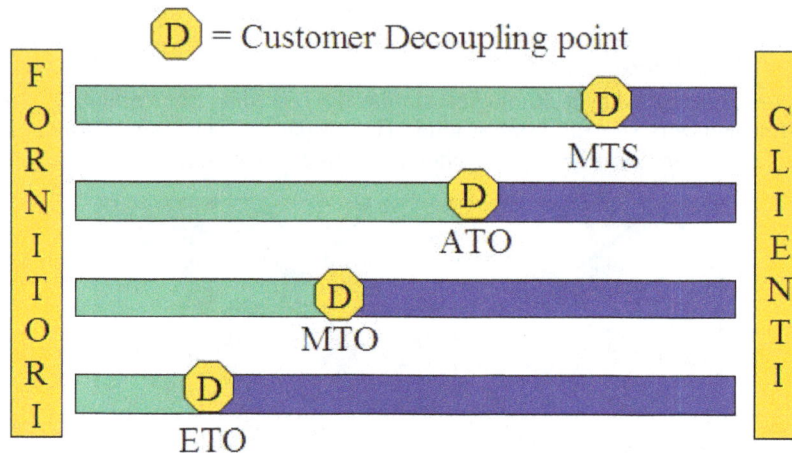

Delivery lead time is the blue bar, manufacturing time is the whole bar, the green bar is the difference between the two.

Another possible classification is one based on Lead Time (manufacturing lead time vs delivery lead time): Engineer to Order (ETO, Purchase to Order (PTO), Make to Order (MTO), Assemble to Order (ATO) and Make to Stock (MTS). According to this classification different kinds of systems will have different customer order decoupling points (CODP), meaning that Work in Progress (WIP) cycle stock levels are practically nonexistent regarding operations located after the CODP (except for WIP due to queues).

The concept of production systems can be expanded to the service sector world keeping in mind that services have some fundamental differences in respect to material goods: intangibility, client always present during transformation processes, no stocks for "finished goods". Services can be classified according to a service process matrix: degree of labor intensity (volume) vs degree of customization (variety). With a high degree of labor intensity there are Mass Services (e.g., commercial banking bill payments and state schools) and Professional Services (e.g., personal physicians and lawyers), while with a low degree of labor intensity there are Service Factories (e.g., airlines and hotels) and Service Shops (e.g., hospitals and auto mechanics).

The systems described above are ideal types: real systems may present themselves as hybrids of those categories. Consider, for example, that the production of jeans involves initially carding, spinning, dyeing and weaving, then cutting the fabric in different shapes and assembling the parts in pants or jackets by combining the fabric with thread, zippers and buttons, finally finishing and distressing the pants/jackets before being shipped to stores. The beginning can be seen as process production, the middle as part production and the end again as process production: it's unlikely that a single company will keep all the stages of production under a single roof, therefore the problem of vertical integration and outsourcing arises. Most products require, *from a supply chain perspective*, both process production and part production.

Metrics: Efficiency and Effectiveness

Operations strategy concerns policies and plans of use of the firm productive resources with the aim of supporting long term competitive strategy. Metrics in operations management can be broadly classified into efficiency metrics and effectiveness metrics. Effectiveness metrics involve:

1. Price (actually fixed by marketing, but lower bounded by production cost): purchase price, use costs, maintenance costs, upgrade costs, disposal costs

2. Quality: specification and compliance

3. Time: productive lead time, information lead time, punctuality

4. Flexibility: mix, volume, gamma

5. Stock availability

6. Ecological Soundness: biological and environmental impacts of the system under study.

A more recent approach, introduced by Terry Hill, involves distinguishing competitive variables in order winner and order qualifiers when defining operations strategy. Order winners are variables which permit differentiating the company from competitors, while order qualifiers are prerequisites for engaging in a transaction. This view can be seen as a unifying approach between operations management and marketing.

Productivity is a standard efficiency metric for evaluation of production systems, broadly speaking a ratio between outputs and inputs, and can assume many specific forms, for example: machine productivity, workforce productivity, raw material productivity, warehouse productivity (=inventory turnover). It is also useful to break up productivity in use U (productive percentage of total time) and yield η (ratio between produced volume and productive time) to better evaluate production systems performances. Cycle times can be modeled through manufacturing engineering if the individual operations are heavily automated, if the manual component is the prevalent one, methods used include: time and motion study, predetermined motion time systems and work sampling.

ABC analysis is a method for analyzing inventory based on Pareto distribution, it posits that since revenue from items on inventory will be power law distributed then it makes sense to manage items differently based on their position on a revenue-inventory level matrix, 3 classes are constructed (A,B and C) from cumulative item revenues, so in a matrix each item will have a letter (A,B or C) assigned for revenue and inventory. This method posits that items away from the diagonal should be managed differently: items in the upper part are subject to risk of obsolescence, items in the lower part are subject to risk of stockout.

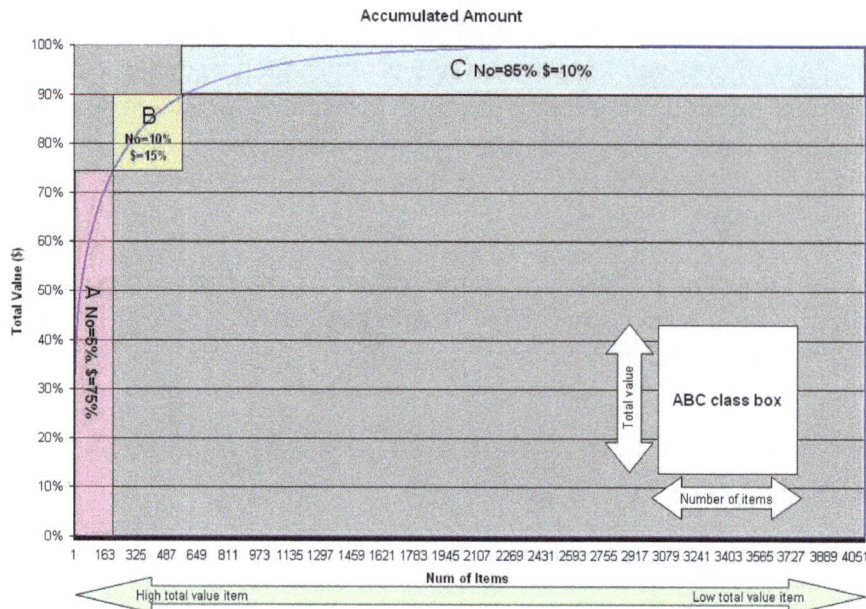

An ABC cumulated curve. Typically one curve is constructed for revenue
(consumption) and another for inventory (stock).

Throughput is a variable which quantifies the number of parts produced in the unit of time. Although estimating throughput for a single process maybe fairly simple, doing so for an entire production system involves an additional difficulty due to the presence of queues which can come from: machine breakdowns, processing time variability, scraps, setups, maintenance time, lack of orders, lack of materials, strikes, bad coordination between resources, mix variability, plus all these inefficiencies tend to compound depending on the nature of the production system. One important example of how system throughput is tied to system design are bottlenecks: in job shops bottlenecks are typically dynamic and dependent on scheduling while on transfer lines it makes sense to speak of "the bottleneck" since it can be univocally associated with a specific station on the line. This leads to the problem of how to define capacity measures, that is an estimation of the maximum output of a given production system, and capacity utilization.

Overall Equipment Effectiveness (OEE) is defined as the product between system availability, cycle time efficiency and quality rate. OEE is typically used as key performance indicator (KPI) in conjunction with the lean manufacturing approach.

Configuration and Management

Designing the *configuration of production systems* involves both technological and organizational variables. Choices in production technology involve: dimensioning capacity, fractioning capacity, capacity location, outsourcing processes, process technology, automation of operations, trade-off between volume and variety. Choices in the organizational area involve: defining worker skills and responsibilities, team coordination, worker incentives and information flow.

Regarding *production planning*, there is a basic distinction between the push approach and the pull approach, with the later including the singular approach of just in time. Pull means that the production system authorizes production based on inventory level; push means that production occurs based on demand (forecasted or present, that is purchase orders). An individual production system can be both push and pull; for example activities before the CODP may work under a pull system, while activities after the CODP may work under a push system.

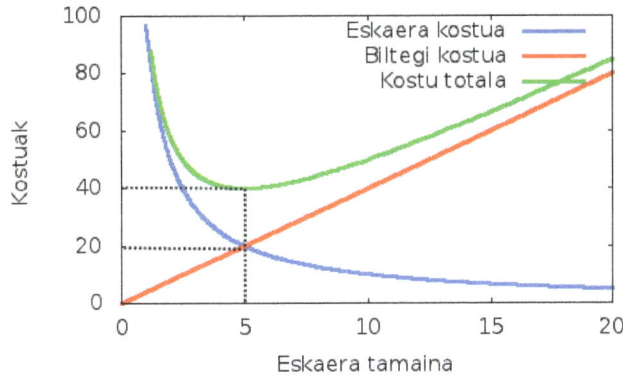

Classic EOQ model: trade-off between ordering cost (blue) and holding cost (red).
Total cost (green) admits a global optimum.

Regarding the traditional pull approach to inventory control, a number of techniques have been developed based on the work of Ford W. Harris (1913), which came to be known as the economic order quantity (EOQ) model. This model marks the beginning of inventory theory, which includes the Wagner-Within Procedure, the News Vendor Model, Base Stock Model and the Fixed Time Period model. These models usually involve the calculation of cycle stocks and buffer stocks, the latter usually modeled as a function of demand variability. The Economic Production Quantity (EPQ) differs from the EOQ model only in that it assumes a constant fill rate for the part being produced, instead of the instantaneous refilling of the EOQ model.

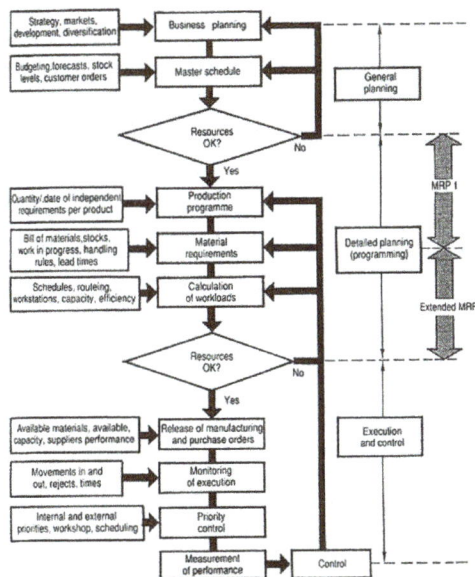

A typical MRPII construct: general planning (top) concerned with forecasts, capacity planning and inventory levels, programming (middle) concerned with calculation of workloads, rough-cut capacity planning, MPS, capacity requirements planning, traditional MRP planning, control (bottom) concerned with scheduling.

Joseph Orlickly and others at IBM developed a push approach to inventory control and production planning, now known as material requirements planning (MRP), which takes as input both the Master Production Schedule (MPS) and the Bill of Materials (BOM) and gives as output a schedule for the materials (components) needed in the production process. MRP therefore is a planning tool to manage purchase orders and production orders (also called jobs).

The MPS can be seen as a kind of aggregate planning for production coming in two fundamentally opposing varieties: plans which try to chase demand and level plans which try to keep uniform capacity utilization. Many models have been proposed to solve MPS problems:

- Analytical models (e.g. Magee Boodman model)

- Exact optimization algorithmic models (e.g. LP and ILP)

- Heuristic models (e.g. Aucamp model).

MRP can be briefly described as a 3s procedure: sum (different orders), split (in lots), shift (in time according to item lead time). To avoid an "explosion" of data processing in MRP (number of BOMs required in input) planning bills (such as family bills or super bills) can be useful since they allow a rationalization of input data into common codes. MRP had some notorious problems such as infinite capacity and fixed lead times, which influenced successive modifications of the original software architecture in the form of MRP II, enterprise resource planning (ERP) and Advanced planning and scheduling (APS).

In this context problems of scheduling (sequencing of production), loading (tools to use), part type selection (parts to work on) and applications of operations research have a significant role to play.

Lean manufacturing is an approach to production which arose in Toyota between the end of World War II and the seventies. It comes mainly from the ideas of Taiichi Ohno and Toyoda Sakichi which are centered on the complementary notions of just in time and autonomation (jidoka), all aimed at reducing waste (usually applied in PDCA style). Some additional elements are also fundamental: production smoothing (Heijunka), capacity buffers, setup reduction, cross-training and plant layout.

- Heijunka: production smoothing presupposes a level strategy for the MPS and a final assembly schedule developed from the MPS by smoothing aggregate production requirements in smaller time buckets and sequencing final assembly to achieve repetitive manufacturing. *If these conditions are met*, expected throughput can be equaled to the inverse of takt time. Besides volume, heijunka also means attaining mixed model production, which however may only be feasible through set-up reduction. A standard tool for achieving this is the Heijunka box.

- Capacity buffers: ideally a JIT system would work with zero breakdowns, this however is very hard to achieve in practice, nonetheless Toyota favors acquiring extra capacity over extra WIP to deal with starvation.

- Set-up reduction: typically necessary to achieve mixed model production, a key distinction can be made between internal and external setup. Internal setups (e.g. removing a die) refers to tasks when the machine is not working, while external setups can be completed while the machine is running (ex:transporting dies).

- Cross training: important as an element of Autonomation, Toyota cross trained their employees through rotation, this served as an element of production flexibility, holistic thinking and reducing boredom.

- Layout: U-shaped lines or cells are common in the lean approach since they allow for minimum walking, greater worker efficiency and flexible capacity.

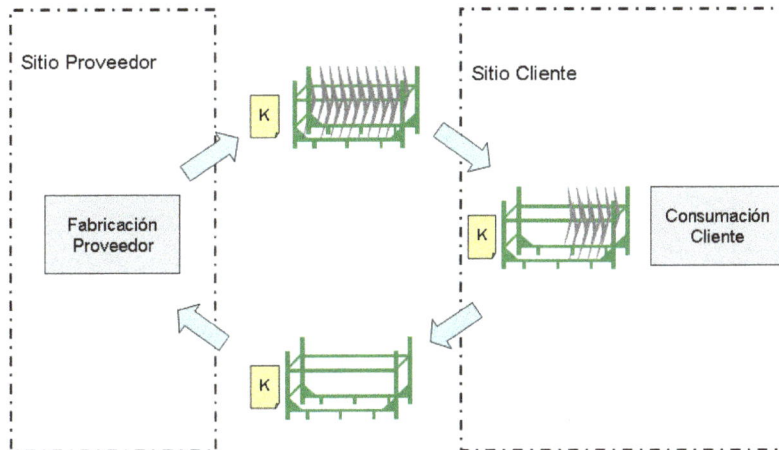

When introducing kanbans in real production systems, attaining unitary lot from the start maybe unfeasible, therefore the kanban will represent a given lot size defined by management.

A series of tools have been developed mainly with the objective of replicating Toyota success: a very common implementation involves small cards known as kanbans; these also come in some varieties: reorder kanbans, alarm kanbans, triangular kanbans, etc. In the classic kanban procedure with one card:

- Parts are kept in containers with their respective kanbans

- The downstream station moves the kanban to the upstream station and starts producing the part at the downstream station

- The upstream operator takes the most urgent kanban from his list (compare to queue discipline from queue theory) and produces it and attach its respective kanban

The two-card kanban procedure differs a bit:

- The downstream operator takes the production kanban from his list

- If required parts are available he removes the move kanban and places them in another box, otherwise he chooses another production card

- He produces the part and attach its respective production kanban

- Periodically a mover picks up the move kanbans in upstream stations and search for the respective parts, when found he exchanges production kanbans for move kanbans and move the parts to downstream stations

Since the number of kanbans in the production system is set by managers as a constant number, the kanban procedure works as WIP controlling device, which for a given arrival rate, per Little's Law, works as a lead time controlling device.

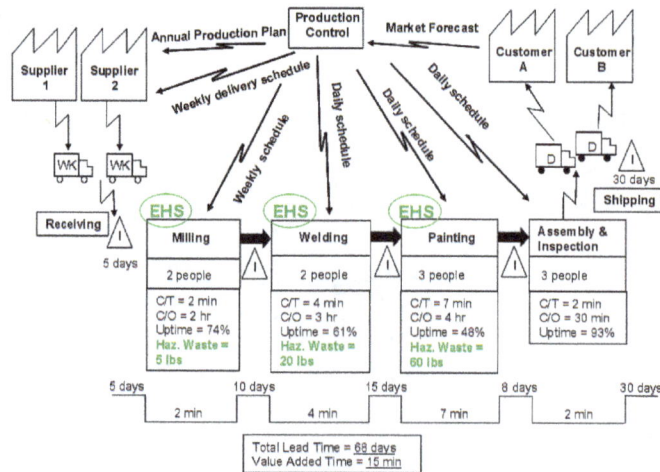

Value Stream Mapping, a representation of materials and information flows inside a company, mainly used in the lean manufacturing approach. The calculation of the time-line (bottom) usually involves using Little's Law to derive lead time from stock levels and takt time.

In Toyota the TPS represented more of a philosophy of production than a set of specific lean tools, the latter would include:

- SMED: a method for reducing changeover times

- Value Stream Mapping: a graphical method for analyzing the current state and designing a future state

- lot-size reduction

- elimination of time batching

- Rank Order Clustering: an algorithm which groups machines and product families together, used for designing manufacturing cells

- single-point scheduling, the opposite of the traditional push approach

- multi-process handling: when one operator is responsible for operating several machines or processes

- poka-yoke: any mechanism in lean manufacturing that helps an equipment operator avoid (*yokeru*) mistakes (*poka*)

- 5S: describes how to organize a work space for efficiency and effectiveness by identifying and storing the items used, maintaining the area and items, and sustaining the new order

- backflush accounting: a product costing approach in which costing is delayed until goods are finished

Seen more broadly, JIT can include methods such as: product standardization and modularity, group technology, total productive maintenance, job enlargement, job enrichment, flat organization and vendor rating (JIT production is very sensitive to replenishment conditions).

In heavily automated production systems production planning and information gathering may be executed via the control system, attention should be paid however to avoid problems suck as deadlocks, as these can lead to productivity losses.

Mathematical Modeling

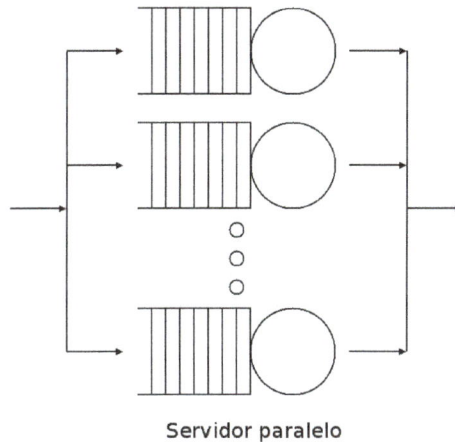

Servidor paralelo

Queue networks are systems in which single queues are connected by a routing network. In this image servers are represented by circles, queues by a series of retangles and the routing network by arrows. In the study of queue networks one typically tries to obtain the equilibrium distribution of the network.

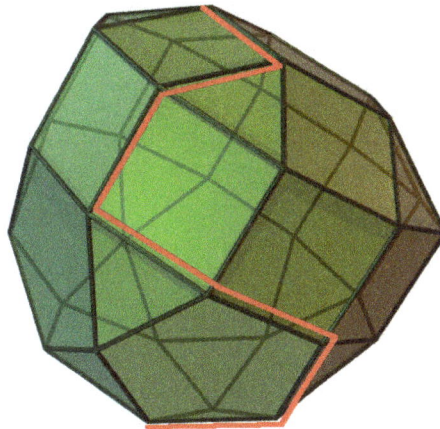

Illustration of the Simplex method, a classical approach to solving LP optimization problems and also integer programming (ex:branch and cut). Mainly used in push approach but also in production system configuration. The interior of the green polytope geometrically represents the feasible region, while the red line indicates the sequence of pivot operations required to reach the optimal solution.

There are also fields of mathematical theory which have found applications in the field of operations management such as operations research: mainly mathematical optimization problems and queue theory. Queue theory is employed in modelling queue and processing times in production systems while mathematical optimization draws heavily from multivariate calculus and linear algebra. Queue theory is based on Markov chains and stochastic processes. It also worth noticing that computations of safety stocks are usually based on modeling demand as a normal distribution and MRP and some inventory problems can be formulated using optimal control.

When analytical models are not enough, managers may resort to using simulation. Simulation has been traditionally done thought the Discrete event simulation paradigm, where the simulation model possesses a state which can only change when a discrete event happens, which consists of a clock and list of events. The more recent Transaction-level modeling paradigm consists of a set of resources and a set of transactions: transactions move through a network of resources (nodes) according to a code, called process.

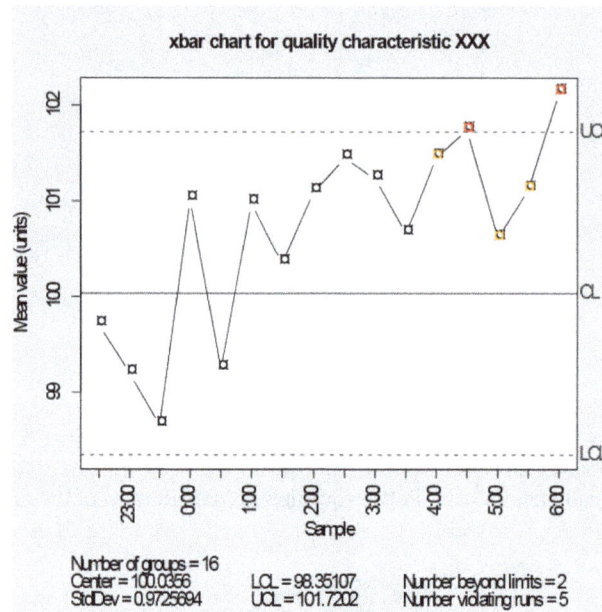

A control chart: process output variable is modeled by a probability density function and for each statistic of the sample an upper control line and lower control line are fixed, when the statistic moves out of bounds, an alarm is given and possible causes are investigated. In this drawing the statistic of choice is the mean and red points represent alarm points.

Since real production processes are always affected by disturbances in both inputs and outputs, many companies implement some form of Quality management or quality control. The Seven Basic Tools of Quality designation provides a summary of commonly used tools:

- check sheets

- Pareto charts

- Ishikawa diagrams (Cause-and-effect diagram)

- control charts

- histogram

- scatter diagram

- stratification

These are used in approaches like Total quality management and Six Sigma. Keeping quality under control is relevant to both increasing customer satisfaction and reducing processing waste.

Operations management textbooks usually cover demand forecasting, even though it is not strictly speaking an operations problem, because demand is related to some production systems variables. For example, a classic approach in dimensioning safety stocks requires calculating standard deviation of forecast errors. Demand forecasting is also a critical part of push systems, since order releases have to be planned ahead of actual clients orders. Also any serious discussion of capacity planning involves adjusting company outputs with market demands.

Safety, Risk and Maintenance

Other important management problems involve maintenance policies, safety management systems, facility management and supply chain integration.

Organizations

The following organizations support and promote operations management:

- Association for Operations Management (APICS) which supports the *Production and Inventory Management Journal*

- European Operations Management Association (EurOMA) which supports the International Journal of Operations & Production Management

- Production and Operations Management Society (POMS) which supports the journal: *Production and Operations Management*

- Institute for Operations Research and the Management Sciences (INFORMS)

- The Manufacturing and Service Operations Management Society (MSOM) which supports the journal: Manufacturing & Service Operations Management

- Institute of Operations Management (UK)

- Association of Technology, Management, and Applied Engineering (ATMAE)

Manufacturing Operations Management

Manufacturing operations management (MOM) is a methodology for viewing an end-to-end manufacturing process with a view to optimizing efficiency.

There are many types of MOM software, including for production management, performance analysis, quality and compliance, and human machine interface (HMI). Production management software provides real-time information about jobs and orders, labor and materials, machine status, and product shipments. Performance analysis software displays metrics at the machine, line, plant and enterprise level for situational or historical analysis. Quality and compliance software is used to promote compliance with standards and specifications for operational processes and procedures. HMI software is a form of manufacturing operations management (MOM) software that enables operators to manage industrial and process control machinery using a computer-based interface.

Emerging Software Trends

Advancements in technology and market demands are enabling new capabilities in MOM software platforms, gradually closing gaps in end-user needs.

- Collaboration Capabilities: Collaboration and workflow services support people-to-people, people-to-systems, and systems-to-systems interactions, enforcing procedures and rules while flexibly adapting to real-time situations with alternate workflows and processes.

- Security Services: Future manufacturing platforms will leverage common security services that determine roles, responsibilities, authorities, and access across all systems and application functions while fitting into corporate IT security schemes.

- Asset & Production Model: Future manufacturing platforms will have a unified asset and production model that supports all of the interrelationships between physical production equipment, facilities, inventory/materials and people, as well as production definitions such as the manufacturing bill of materials, productions orders, etc. This contrasts with older systems that either had subsets of these interrelationships across multiple databases, or could not effectively deal with federating across multiple systems of record.

- Operations Database & Historians: Evolving from older systems that had separate historians and production databases that were difficult to correlate across, service-based platforms will have a unified operations database and historian. This will capture and aggregate all time-series and production event information surrounding everything involved in each product and production run with a full genealogy of components and materials, related performance information, and federation across other systems and devices of record.

- Visualization and Mobility: Today, different MOM applications support different graphical user interfaces, Web interfaces, specific mobile applications, etc. The future manufacturing platform will provide common visualization and mobility for a consistent user interface experience across different form factors, supporting dedicated and mobile workers that are orchestrated by consistent workflows and procedures.

- Smaller and Focused 'Apps': Today's monolithic systems and applications have too many interdependencies of databases, operate inconsistently, and are not inherently integrated. Being able to take advantage of many of the common software platform services described above, modular apps will be significantly smaller, simpler, and focused. These apps will be much lighter weight in functionality, and, as a result, significantly easier and faster to develop.

Manufacturing Operations Management (MOM) Software Solution Providers

- http://pinpointinfo.com/ Manufacturing Operations Management (MOM) / Manufacturing Execution System (MES) software.

- LYNQ provides web enabled scheduling, dispatching, execution, data collection, tracking and analysis for small to midsize manufacturers

- Inductive Automation (Makers of Ignition)

- SAP

- Innovar Systems Smart Manufacturing

- Apache OFBiz Manufacturing

- Atachi NGIMES

- Dassault Systemes Apriso

- Lighthouse Systems

- LeaderMES

- General Electric Intelligent Platforms

- Wonderware Manufacturing Operations Management

- Queris MES

- Rockwell Automation Suite

- Siemens Camstar Enterprise Platform

- Siemens MOM Headquarters in Genoa -Italy

- Siemens MES/MOM

- Simio Simulation and Scheduling Software - provides simulation based scheduling to produce the most optimal schedule.

- PREACTOR/SIEMENS: PREACTOR (www.preactor.com) is a SIEMENS product for Planning and Scheduling

- Aptean - Factory MES

- Odoo Warehouse

- Talika PMS

- CyberPlan

- AIMMS

- EMANS/ANASOFT MOM for Zero Fault Production, Perfection in Assembly, and Traceability.

- OSIsoft/PI System is an open infrastructure to connect sensor-based data, operations and people to enable real-time intelligence

- Plex Systems Plex Systems is a cloud-based manufacturing ERP solution with a complete MES/MOM solution.

- Procurify Procurify is a cloud-based manufacturing solution with a focus on the SMB market.

- iBASEt Solumina

- Digital Purchase Order Cloud-based e-procurement software solution available for desktop and as mobile app.

- Manufacturing Intelligence Standards and cloud-based MOM software supporting traditional manufacturing, agriculture, oil and gas and mining industries.

References

- Malakooti, Behnam (2013). Operations and Production Systems with Multiple Objectives. John Wiley & Sons. ISBN 978-1-118-58537-5.

- Shewhart, Walter Andrew, Economic control of quality of manufactured product, 1931, New York: D. Van Nostrand Company. pp. 501 p.. ISBN 0-87389-076-0 (edition 1st). LCCN 132090. OCLC 1045408. LCC TS155 .S47.

- Harris, Ford W. (1990) [Reprint from 1913]. "How Many Parts to Make at Once" (PDF). Operations Research. INFORMS. 38 (6): 947–950. doi:10.1287/opre.38.6.947. JSTOR 170962. Retrieved Nov 21, 2012.

- "OR / Pubs / IOL Home". INFORMS.org. 2 January 2009. Archived from the original on 27 May 2009. Retrieved 13 November 2011.

- "operations research (industrial engineering) :: History – Britannica Online Encyclopedia". Britannica.com. Retrieved 13 November 2011.

- ""Numbers are Essential": Victory in the North Atlantic Reconsidered, March–May 1943". Familyheritage.ca. 24 May 1943. Retrieved 13 November 2011.

- "International Journal of Operations Research and Information Systems (IJORIS) (1947–9328)(1947–9336): wJohn Wang: Journals". IGI Global. Retrieved 13 November 2011.

Product: Design, Planning and Development

One of the most important facets of industrial engineering is the product. The following chapter will provide an insightful account on the various processes through which a product takes its final shape. Processes like industrial design, production planning, engineering design process and quality assurance are described thoroughly in the following chapter.

Industrial Design

Industrial design is a process of design applied to products that are to be manufactured through techniques of mass production. Its key characteristic is that design is separated from manufacture: the creative act of determining and defining a product's form and features takes place in advance of the physical act of making a product, which consists purely of repeated, often automated, replication. This distinguishes industrial design from craft-based design, where the form of the product is determined by the product's creator at the time of its creation.

An iPod, an industrially designed product.

All manufactured products are the result of a design process, but the nature of this process can take many forms: it can be conducted by an individual or a large team; it can emphasize intuitive creativity or calculated scientific decision-making, and often emphasizes both at the same time; and it can be influenced by factors as varied as materials, production processes, business strategy and prevailing social, commercial or aesthetic attitudes. The role of an industrial designer is to create and execute design solutions for problems of form, function, usability, physical ergonomics, marketing, brand development, and sales.

KitchenAid 5 qt. Stand Mixer, designed in 1937 by Egmont Arens, remains very successful today

Western Electric Model 302 telephone, found throughout the United States from 1937 until the introduction of touch-tone dialing.

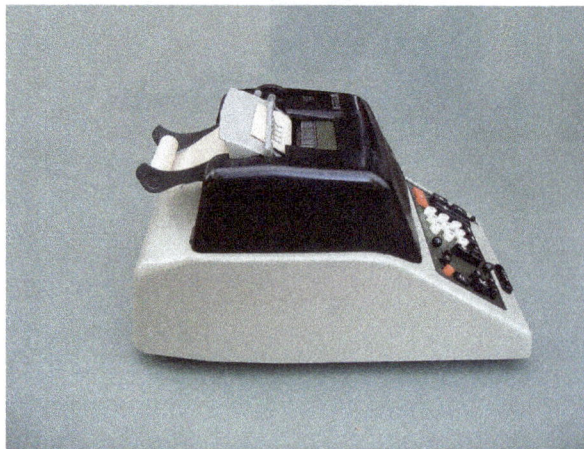

Calculator Olivetti Divisumma 24 designed in 1956 by Marcello Nizzoli

History

Precursors

For several millennia before the onset of industrialisation design, technical expertise and manufacture lay together in the hands of individual craftsmen, who determined the form of a product at the point of its creation according to their own manual skill, the parameters set by their clients, the experience accumulated through their own experimentation and traditional knowledge passed on to them through training or apprenticeship.

The division of labour that underlies the practice of industrial design did have precedents in the pre-industrial era. The growth of trade in the medieval period led to the emergence of large workshops in cities such as Florence, Venice, Nuremberg and Bruges, where groups of more specialist craftsmen made objects with common forms through the repetitive duplication of models defined by their shared training and technique. Competitive pressures in the early 16th century led to the emergence in Italy and Germany of pattern books: collections of engravings illustrating decorative forms and motifs that could be applied to a wide range of products, and whose creation therefore took place in advance of their application. The use of drawing to specify how something is later to be constructed was first developed by architects and shipwrights during the Italian Renaissance.

In the 17th century the scale of the artistic patronage of centralised monarchical states such as France led to the growth of large government-operated manufacturing operations epitomised by the Gobelins Manufactory, opened in Paris in 1667 by Louis XIV. Here teams of hundreds of craftsmen, including specialist artists, decorators and engravers, produced sumptuously decorated products ranging from tapestries and furniture to metalwork and coaches, all under the creative supervision of the King's leading artist Charles Le Brun. This pattern of large-scale royal patronage was repeated in the court porcelain factories of the early 18th century, such as the Meissen porcelain workshops established in 1709 by the Grand Duke of Saxony, where patterns from a range of sources, including court goldsmiths, sculptors and engravers, were used as models for the vessels and figurines for which it became famous. As long as reproduction remained craft-based, however, the form and artistic quality of the product remained in the hands of the individual craftsman, and tended to decline as the scale of production increased.

Birth of Industrial Design

The emergence of industrial design is specifically linked to the growth of industrialisation and mechanisation that began with the industrial revolution in Great Britain in the mid 18th century. The rise of industrial manufacture changed the way objects were made, urbanisation changed patterns of consumption, the growth of empires broadened tastes and diversified markets, and the emergence of a wider middle class created demand for fashionable styles from a much larger and more heterogeneous population.

The first use of the term "industrial design" is often attributed to the industrial designer Joseph Claude Sinel in 1919 (although he himself denied this in interviews), but the discipline predates 1919 by at least a decade. Christopher Dresser is considered among the first independent industrial designers. Industrial design's origins lie in the industrialization of consumer products.

For instance the Deutscher Werkbund, founded in 1907 and a precursor to the Bauhaus, was a state-sponsored effort to integrate traditional crafts and industrial mass-production techniques, to put Germany on a competitive footing with England and the United States.

The earliest use of the term may have been in *The Art Union*, A monthly Journal of the Fine Arts, 1839.

Dyce's report to the Board of Trade on foreign schools of Design for Manufactures. Mr Dyces official visit to France, Prussia and Bavaria for the purpose of examining the state of schools of design in those countries will be fresh in the recollection of our readers. His report on this subject was ordered to be printed some few months since, on the motion of Mr Hume.

The school of St Peter, at Lyons was founded about 1750 for the instruction of draftsmen employed in preparing patterns for the silk manufacture. It has been much more successful than the Paris school and having been disorganized by the revolution, was restored by Napoleon and differently constituted, being then erected into an Academy of Fine Art: to which the study of design for silk manufacture was merely attached as a subordinate branch. It appears that all the students who entered the school commence as if they were intended for artists in the higher sense of the word and are not expected to decide as to whether they will devote themselves to the Fine Arts or to Industrial Design, until they have completed their exercises in drawing and painting of the figure from the antique and from the living model. It is for this reason, and from the fact that artists for industrial purposes are both well paid and highly considered (as being well instructed men) that so many individuals in France engage themselves in *both* pursuits.

The Practical Draughtsman's Book of Industrial Design by Jacques-Eugène Armengaud was printed in 1853. The subtitle of the (translated) work explains, that it wants to offer a "complete course of mechanical, engineering, and architectural drawing." The study of those types of technical drawing, according to Armengaud, belong to the field of industrial design. This work paved the way for a big expansion in the field drawing education in France, the United Kingdom and the United States.

Robert Lepper helped to establish one of the country's first industrial design degree programs at Carnegie Institute of Technology.

Education

Product design and industrial design overlap in the fields of user interface design, information design, and interaction design. Various schools of industrial design specialize in one of these aspects, ranging from pure art colleges (product styling), to mixed programs of engineering and design, to related disciplines such as exhibit design and interior design, to schools that almost completely subordinated aesthetic design to concerns of usage and ergonomics, the so-called *functionalist* school. Except for certain functional areas of overlap between industrial design and engineering design, educational programs in the U.S. for engineering design require accreditation by the Accreditation Board for Engineering and Technology (ABET) in contrast to programs for industrial design which are accredited by the National Association of Schools of Art and Design (NASAD).

Institutions

Most industrial designers complete a design or related program at a vocational school or university. Relevant programs include graphic design, interior design, industrial design, architectural technology, and drafting Diplomas and degrees in industrial design are offered at vocational schools and universities worldwide. Diplomas and degrees take two to four years of study. The study results in a Bachelor of Industrial Design (B.I.D.), Bachelor of Science (B.Sc) or Bachelor of Fine Arts (B.F.A.). Afterwards, the bachelor programme can be extended to postgraduate degrees such as Master of Design, Master of Fine Arts and others to a Master of Arts or Master of Science.

Definition of Industrial Design

Industrial design studies function and form—and the connection between product, user, and environment. Generally, industrial design professionals work in small scale design, rather than overall design of complex systems such as buildings or ships. Industrial designers don't usually design motors, electrical circuits, or gearing that make machines move, but they may affect technical aspects through usability design and form relationships. Usually, they work with other professionals such as engineers who design the mechanical aspects of the product assuring functionality and manufacturability, and with marketers to identify and fulfill customer needs and expectations.

Industrial design (ID) is the professional service of creating and developing concepts and specifications that optimize the function, value and appearance of products and systems for the mutual benefit of both user and manufacturer.

Design, itself, is often difficult to describe to non-designers and engineers, because the meaning accepted by the design community is not made of words. Instead, the definition is created as a result of acquiring a critical framework for the analysis and creation of artifacts. One of the many accepted (but intentionally unspecific) definitions of design originates from Carnegie Mellon's School of Design, "Design is the process of taking something from its existing state and moving it to a preferred state." This applies to new artifacts, whose existing state is undefined, and previously created artifacts, whose state stands to be improved.

Industrial design can overlap significantly with engineering design, and in different countries the boundaries of the two concepts can vary, but in general engineering focuses principally on functionality or Utility of Products whereas industrial design focuses principally on *aesthetic and user-interface* aspects of products. In many jurisdictions this distinction is effectively defined by credentials and/or licensure required to engage in the practice of engineering. "Industrial design" as such does not overlap much with the engineering sub-discipline of industrial engineering, except for the latter's sub-specialty of ergonomics.

At the 29th General Assembly in Gwangju, South Korea,2015, the Professional Practise Committee unveiled a renewed definition of industrial design as follows: "Industrial Design is a strategic problem-solving process that drives innovation, builds business success and leads to a better quality of life through innovative products, systems, services and experiences." An extended version of this definition is as follows: "Industrial Design is a strategic problem-solving process that drives innovation, builds business success and leads to a better quality of life through innovative products, systems, services and experiences. Industrial Design bridges the gap between what is and what's possible. It is a trans-disciplinary profession that harnesses creativity to resolve problems and co-create solutions with the intent of making a product, system, service, experience or a business, better. At its heart, Industrial Design provides a more optimistic way of looking at the future by reframing problems as opportunities. It links innovation, technology, research, business and customers to provide new value and competitive advantage across economic, social and environmental spheres. Industrial Designers place the human in the centre of the process. They acquire a deep understanding of user needs through empathy and apply a pragmatic, user centric problem solving process to design products, systems, services and experiences. They are strategic stakeholders in the innovation process and are uniquely positioned to bridge varied professional disciplines and business interests. They value the economic, social and environmental impact of their work and their contribution towards co-creating a better quality of life."

Design Process

A Fender Stratocaster with sunburst finish, one of the most widely recognized electric guitars in the world.

Model 1300 Volkswagen Beetle

Although the process of design may be considered 'creative,' many analytical processes also take place. In fact, many industrial designers often use various design methodologies in their creative process. Some of the processes that are commonly used are user research, sketching, comparative product research, model making, prototyping and testing. These processes are best defined by the industrial designers and/or other team members. Industrial designers often utilize 3D software, computer-aided industrial design and CAD programs to move from concept to production. They may also build a prototype first and then use industrial CT scanning to test for interior defects and generate a CAD model. From this the manufacturing process may be modified to improve the product.

Product characteristics specified by industrial designers may include the overall form of the object, the location of details with respect to one another, colors, texture, form, and aspects concerning the use of the product. Additionally they may specify aspects concerning the production process, choice of materials and the way the product is presented to the consumer at the point of sale. The inclusion of industrial designers in a product development process may lead to added value by improving usability, lowering production costs and developing more appealing products.

Industrial design may also focus on technical concepts, products, and processes. In addition to aesthetics, usability, and ergonomics, it can also encompass engineering, usefulness, market placement, and other concerns—such as psychology, desire, and the emotional attachment of the user. These values and accompanying aspects that form the basis of industrial design can vary—between different schools of thought, and among practicing designers.

Industrial Design Rights

Industrial design rights are intellectual property rights that make exclusive the visual design of objects that are not purely utilitarian. A design patent would also be considered under this category. An industrial design consists of the creation of a shape, configuration or composition of pattern or color, or combination of pattern and color in three-dimensional form containing aesthetic value. An industrial design can be a two- or three-dimensional pattern used to produce a product, industrial commodity or handicraft. Under the Hague Agreement Concerning the International Deposit of Industrial Designs, a WIPO-administered treaty, a procedure for an international registration exists. An applicant can file for a single international deposit with WIPO or with the national office in a country party to the treaty. The design will then be protected in as many member countries of the treaty as desired.

Examples of Iconic Industrial Design

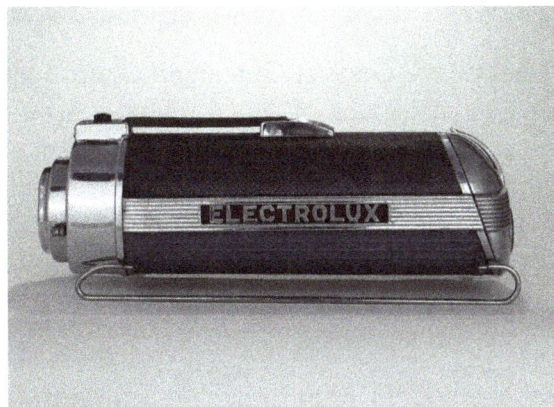

Lurelle Guild. Vacuum Cleaner, ca. 1937. Brooklyn Museum

Chair by Charles Eames

Russel Wright. Coffee Urn, ca. 1935 Brooklyn Museum

A number of industrial designers have made such a significant impact on culture and daily life that their work is documented by historians of social science. Alvar Aalto, renowned as an architect, also designed a significant number of household items, such as chairs, stools, lamps, a tea-cart, and vases. Raymond Loewy was a prolific American designer who is responsible for the Royal Dutch Shell corporate logo, the original BP logo (in use until 2000), the PRR S1 steam locomotive, the Studebaker Starlight (including the later iconic bulletnose), as well as Schick electric razors, Electrolux refrigerators, short-wave radios, Le Creuset French ovens, and a complete line of modern furniture, among many other items.

Richard A. Teague, who spent most of his career with the American Motor Company, originated the concept of using interchangeable body panels so as to create a wide array of different vehicles using the same stampings. He was responsible for such unique automotive designs as the Pacer, Gremlin, Matador coupe, Jeep Cherokee, and the complete interior of the Eagle Premier.

Viktor Schreckengost designed bicycles manufactured by Murray bicycles for Murray and Sears, Roebuck and Company. With engineer Ray Spiller, he designed the first truck with a cab-over-engine configuration, a design in use to this day. Schreckengost also founded The Cleveland Institute of Art's school of industrial design.

Oskar Barnack was a German optical engineer, precision mechanic, industrial designer, and the father of 35mm photography. He developed the Leica, which became the hallmark for photography for 50 years, and remains a high-water mark for mechanical and optical design.

Charles and Ray Eames were most famous for their pioneering furniture designs, such as the Eames Lounge Chair Wood and Eames Lounge Chair. Other influential designers included Henry Dreyfuss, Eliot Noyes, John Vassos, and Russel Wright.

Dieter Rams is a German industrial designer closely associated with the consumer products company Braun and the Functionalist school of industrial design.

Many of Apple's recent iconic products were designed by Sir Jonathan Ive.

Production Planning

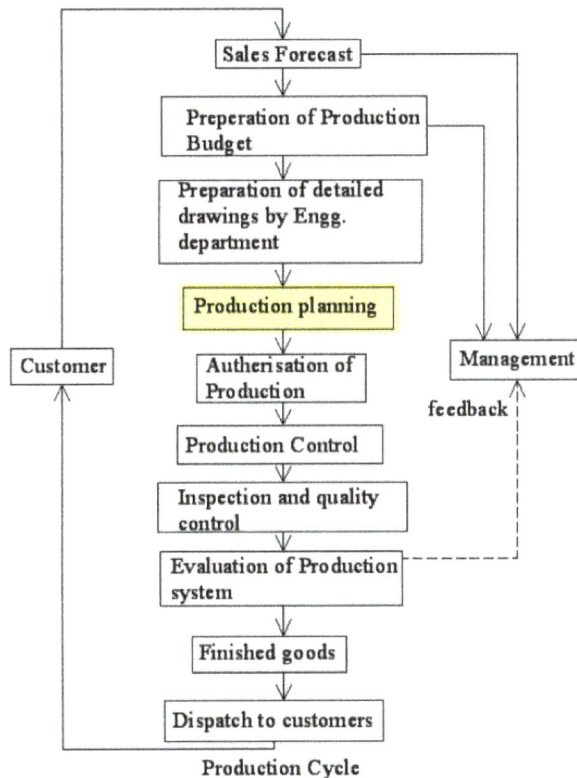

Role of Production Planning in the Production Cycle.

Production planning is the planning of production and manufacturing modules in a company or industry. It utilizes the resource allocation of activities of employees, materials and production capacity, in order to serve different customers.

Different types of production methods, such as single item manufacturing, batch production, mass production, continuous production etc. have their own type of production planning. Production planning can be combined with production control into production planning and control, or it can be combined and or integrated into enterprise resource planning.

Production planning is used in companies in several different industries, including agriculture, industry, amusement industry, etc.

Overview

Production planning is a plan for the future production, in which the facilities needed are determined and arranged. A production planning is made periodically for a specific time period, called the planning horizon. It can comprise the following activities:

- Determination of the required product mix and factory load to satisfy customers needs.

- Matching the required level of production to the existing resources.

- Scheduling and choosing the actual work to be started in the manufacturing facility"

- Setting up and delivering production orders to production facilities.

In order to develop production plans, the production planner or production planning department needs to work closely together with the marketing department and sales department. They can provide sales forecasts, or a listing of customer orders." The "work is usually selected from a variety of product types which may require different resources and serve different customers. Therefore, the selection must optimize customer-independent performance measures such as cycle time and customer-dependent performance measures such as on-time delivery."

A critical factor in production planning is "the accurate estimation of the productive capacity of available resources, yet this is one of the most difficult tasks to perform well." Production planning should always take "into account material availability, resource availability and knowledge of future demand."

History

Modern production planning methods and tools have been developed since late 19th century. Under Scientific Management, the work for each man or each machine is mapped out in advance (see image). The origin of production planning back goes another century. Kaplan (1986) summarized that "the demand for information for internal planning and control apparently arose in the first half of the 19th century when firms, such as textile mills and railroads, had to devise internal administrative procedures to coordinate the multiple processes involved in the performance of the basic activity (the conversion of raw materials into finished goods by textile mills, the transportation of passengers and freight by the railroads."

Herrmann (1996) further describes the circumstances in which new methods for internal planning and control evolved: "The first factories were quite simple and relatively small.

They produced a small number of products in large batches. Productivity gains came from using interchangeable parts to eliminate time-consuming fitting operations. Through the late 1800s, manufacturing firms were concerned with maximizing the productivity of the expensive equipment in the factory. Keeping utilization high was an important objective. Foremen ruled their shops, coordinating all of the activities needed for the limited number of products for which they were responsible. They hired operators, purchased materials, managed production, and delivered the product. They were experts with superior technical skills, and they (not a separate staff of clerks) planned production. Even as factories grew, they were just bigger, not more complex.

Planning department bulletin board, 1911.

About production planning Herrmann (1996) recounts that "production scheduling started simply also. Schedules, when used at all, listed only when work on an order should begin or when the order is due. They didn't provide any information about how long the total order should take or about the time required for individual operations ..."

In 1923 *Industrial Management* cited a Mr. Owens who had observed: "Production planning is rapidly becoming one of the most vital necessities of management. It is true that every establishment, no matter how large or how small has production planning in some form ; but a large percentage of these do not have planning that makes for an even flow of material, and a minimum amount of money tied up in inventories."

Production Planning Topics

Types of Planning

Different types of production planning can be applied:

- Advanced planning and scheduling

- Capacity planning

- Master production schedule

- Material requirements planning

- MRP II

- Scheduling

- Workflow

Related kind of planning in organizations

- Employee scheduling

- Enterprise resource planning

- Inventory control

- Product planning

- Project planning

- Sales and operations planning

- Strategy

Production Control

Production control is the activity of controlling the workflow in the production. It is partly complementary to production planning.

Engineering Design Process

The engineering design process is a methodical series of steps that engineers use in creating functional products and processes. The process is highly iterative - parts of the process often need to be repeated many times before another can be entered - though the part(s) that get iterated and the number of such cycles in any given project can be highly variable.

...It is a decision making process (often iterative) in which the basic sciences, mathematics, and engineering sciences are applied to convert resources optimally to meet a stated objective. Among the fundamental elements of the design process are the establishment of objectives and criteria, synthesis, analysis, construction, testing and evaluation

—ABET

The steps tend to get articulated, subdivided, and/or illustrated in a variety of different ways, but they generally reflect certain core principles regarding the underlying concepts and their respective sequence and interrelationship.

Common Stages of the Engineering Design Process

One framing of the engineering design process delineates the following stages: *research, con-*

ceptualization, feasibility assessment, establishing design requirements, preliminary design, detailed design, production planning and tool design, and production. Others, noting that "different authors (in both research literature and in textbooks) define different phases of the design process with varying activities occurring within them," have suggested more simplified/generalized models - such as *problem definition, conceptual design, preliminary design, detailed design, and design communication.* In both of these examples, other key aspects - such as concept evaluation and prototyping - are subsets and/or extensions of one or more of the listed steps. It's also important to understand that in these as well as other articulations of the process, different terminology employed may have varying degrees of overlap, which affects what steps get stated explicitly or deemed "high level" versus subordinate in any given model.

Research

Various stages of the design process (and even earlier) can involve a significant amount of time spent on locating information and research. Consideration should be given to the existing applicable literature, problems and successes associated with existing solutions, costs, and marketplace needs.

The source of information should be relevant, including existing solutions. Reverse engineering can be an effective technique if other solutions are available on the market. Other sources of information include the Internet, local libraries, available government documents, personal organizations, trade journals, vendor catalogs and individual experts available.

Design Requirements

Establishing design requirements, sometimes termed problem definition, is one of the most important elements in the design process, and this task is often performed at the same time as a feasibility analysis. The design requirements control the design of the project throughout the engineering design process. These include basic things like the functions, attributes, and specifications - determined after assessing user needs. Some design requirements include hardware and software parameters, maintainability, availability, and testability.

Feasibility

In some cases, a feasibility study is carried out after which schedules, resource plans and, estimates for the next phase are developed. The feasibility study is an evaluation and analysis of the potential of a proposed project to support the process of decision making. It outlines and analyses alternatives or methods of achieving the desired outcome. The feasibility study helps to narrow the scope of the project to identify the best scenario. A feasibility report is generated following which Post Feasibility Review is performed.

The purpose of a feasibility assessment is to determine whether the engineer's project can proceed into the design phase. This is based on two criteria: the project needs to be based on an achievable idea, and it needs to be within cost constraints. It is important to have engineers with experience and good judgment to be involved in this portion of the feasibility study.

Conceptualization

A concept study (conceptualization, conceptual engineering) is often a phase of project planning that includes producing ideas and taking into account the pros and cons of implementing those ideas. This stage of a project is done to minimize the likelihood of error, manage costs, assess risks, and evaluate the potential success of the intended project. In any event, once an engineering issue or problem is defined, potential solutions must be identified. These solutions can be found by using ideation, the mental process by which ideas are generated. In fact, this step is often termed Ideation or "Concept Generation." The following are widely used techniques:

- trigger word - a word or phrase associated with the issue at hand is stated, and subsequent words and phrases are evoked.

- morphological chart - independent design characteristics are listed in a chart, and different engineering solutions are proposed for each solution. Normally, a preliminary sketch and short report accompany the morphological chart.

- synectics - the engineer imagines him or herself as the item and asks, "What would I do if I were the system?" This unconventional method of thinking may find a solution to the problem at hand. The vital aspects of the conceptualization step is synthesis. Synthesis is the process of taking the element of the concept and arranging them in the proper way. Synthesis creative process is present in every design.

- brainstorming - this popular method involves thinking of different ideas, typically as part of a small group, and adopting these ideas in some form as a solution to the problem

Various generated ideas must then undergo a concept evaluation step, which utilizes various tools to compare and contrast the relative strengths and weakness of possible alternatives.

Preliminary Design

The preliminary design, or high-level design (also called FEED), often bridges a gap between design conception and detailed design, particularly in cases where the level of conceptualization achieved during ideation is not sufficient for full evaluation. So in this task, the overall system configuration is defined, and schematics, diagrams, and layouts of the project may provide early project configuration. (This notably varies a lot by field, industry, and product.) During detailed design and optimization, the parameters of the part being created will change, but the preliminary design focuses on creating the general framework to build the project on.

Detailed Design

Following FEED is the Detailed Design (Detailed Engineering) phase, which may consist of procurement of materials as well. This phase further elaborates each aspect of the project/product by complete description through solid modeling, drawings as well as specifications.

Some example specifications to be finalized may include:

- Operating parameters

- Operating and nonoperating environmental stimuli

- Test requirements

- External dimensions

- Maintenance and testability provisions

- Materials requirements

- Reliability requirements

- External surface treatment

- Design life

- Packaging requirements

- External marking

Computer-aided design (CAD) programs have made the detailed design phase more efficient. For example, a CAD program can provide optimization to reduce volume without hindering a part's quality. It can also calculate stress and displacement using the finite element method to determine stresses throughout the part.

Production Planning

The production planning and tool design consists of planning how to mass-produce the product and which tools should be used in the manufacturing. Tasks to complete in this step include selecting materials, selection of the production processes, determination of the sequence of operations, and selection of tools such as jigs, fixtures, metal cutting and metal forming tools. This task also involves additional prototype testing iterations to ensure the mass-produced version meets qualification testing standards.

Comparison with the Scientific Method

The engineering design process bears some similarity to the scientific method. Both processes begin with existing knowledge, and gradually become more specific in the search for *knowledge* (in the case of "pure" or basic science) or a *solution* (in the case of "applied" science, such as engineering).

Quality Assurance

Quality assurance (QA) is a way of preventing mistakes or defects in manufactured products and avoiding problems when delivering solutions or services to customers; which ISO 9000 defines as "part of quality management focused on providing confidence that quality requirements will be fulfilled". This defect prevention in quality assurance differs subtly from defect detection and rejection in quality control, and has been referred to as a *shift left* as it focuses on quality earlier in the process.

Quality assurance comprises administrative and procedural activities implemented in a quality system so that requirements and goals for a product, service or activity will be fulfilled. It is the systematic measurement, comparison with a standard, monitoring of processes and an associated feedback loop that confers error prevention. This can be contrasted with quality control, which is focused on process output.

Two principles included in quality assurance are: "Fit for purpose" (the product should be suitable for the intended purpose); and "right first time" (mistakes should be eliminated). QA includes management of the quality of raw materials, assemblies, products and components, services related to production, and management, production and inspection processes.

Suitable quality is determined by product users, clients or customers, not by society in general. It is not related to cost, and adjectives or descriptors such as "high" and "poor" are not applicable. For example, a low priced product may be viewed as having high quality because it is disposable, whereas another may be viewed as having poor quality because it is not disposable.

History

Initial Efforts to Control the Quality of Production

During the Middle Ages, guilds adopted responsibility for the quality of goods and services offered by their members, setting and maintaining certain standards for guild membership.

Royal governments purchasing material were interested in quality control as customers. For this reason, King John of England appointed William Wrotham to report about the construction and repair of ships. Centuries later, Samuel Pepys, Secretary to the British Admiralty, appointed multiple such overseers.

Prior to the extensive division of labor and mechanization resulting from the Industrial Revolution, it was possible for workers to control the quality of their own products. The Industrial Revolution led to a system in which large groups of people performing a specialized type of work were grouped together under the supervision of a foreman who was appointed to control the quality of work manufactured.

Wartime Production

During the time of the First World War, manufacturing processes typically became more complex with larger numbers of workers being supervised. This period saw the widespread introduction of mass production and piece work, which created problems as workmen could now earn more money by the production of extra products, which in turn occasionally led to poor quality workmanship being passed on to the assembly lines. To counter bad workmanship, full-time inspectors were introduced to identify, quarantine and ideally correct product quality failures. Quality control by inspection in the 1920s and 1930s led to the growth of quality inspection functions, separately organized from production and large enough to be headed by superintendents.

The systematic approach to quality started in industrial manufacturing during the 1930s, mostly in the U.S., when some attention was given to the cost of scrap and rework. The impact of mass production required during the Second World War made it necessary to introduce an improved form of quality control known as Statistical Quality Control, or SQC. Some of the initial work for SQC is credited to Walter A. Shewhart of Bell Labs, starting with his famous one-page memorandum of 1924..

SQC includes the concept that every production piece cannot be fully inspected into acceptable and non acceptable batches. By extending the inspection phase and making inspection organizations more efficient, it provides inspectors with control tools such as sampling and control charts, even where 100 percent inspection is not practicable. Standard statistical techniques allow the producer to sample and test a certain proportion of the products for quality to achieve the desired level of confidence in the quality of the entire batch or production run.

Postwar

In the period following World War II, many countries' manufacturing capabilities that had been destroyed during the war were rebuilt. General Douglas MacArthur oversaw the re-building of Japan. During this time, General MacArthur involved two key individuals in the development of modern quality concepts: W. Edwards Deming and Joseph Juran. Both individuals promoted the collaborative concepts of quality to Japanese business and technical groups, and these groups utilized these concepts in the redevelopment of the Japanese economy.

Although there were many individuals trying to lead United States industries towards a more comprehensive approach to quality, the U.S. continued to apply the Quality Control (QC) concepts of inspection and sampling to remove defective product from production lines, essentially ignoring advances in QA for decades.

Approaches

Failure Testing

A valuable process to perform on a whole consumer product is failure testing or stress testing. In mechanical terms this is the operation of a product until it fails, often under stresses such as increasing vibration, temperature, and humidity. This exposes many unanticipated weaknesses in a product, and the data is used to drive engineering and manufacturing process improvements. Often quite simple changes can dramatically improve product service, such as changing to mold-resistant paint or adding lock-washer placement to the training for new assembly personnel.

Statistical Control

Statistical control is based on analyses of objective and subjective data. Many organizations use statistical process control as a tool in any quality improvement effort to track quality data. Any product can be statistically charted as long as they have a common cause variance or special cause variance to track.

Walter Shewart of Bell Telephone Laboratories recognized that when a product is made, data can be taken from scrutinized areas of a sample lot of the part and statistical variances are then analyzed and charted. Control can then be implemented on the part in the form of rework or scrap, or control can be implemented on the process that made the part, ideally eliminating the defect before more parts can be made like it.

Total Quality Management

The quality of products is dependent upon that of the participating constituents, some of which are sustainable and effectively controlled while others are not. The process(es) which are managed with QA pertain to Total Quality Management.

If the specification does not reflect the true quality requirements, the product's quality cannot be guaranteed. For instance, the parameters for a pressure vessel should cover not only the material and dimensions but operating, environmental, safety, reliability and maintainability requirements.

Models and Standards

ISO 17025 is an international standard that specifies the general requirements for the competence to carry out tests and or calibrations. There are 15 management requirements and 10 technical requirements. These requirements outline what a laboratory must do to become accredited. Management system refers to the organization's structure for managing its processes or activities that transform inputs of resources into a product or service which meets the organization's objectives, such as satisfying the customer's quality requirements, complying with regulations, or meeting environmental objectives.

The Capability Maturity Model Integration (CMMI) model is widely used to implement Process and Product Quality Assurance (PPQA) in an organization. The CMMI maturity levels can be divided into 5 steps, which a company can achieve by performing specific activities within the organization.

Company Quality

During the 1980s, the concept of "company quality" with the focus on management and people came to the fore. It was realized that, if all departments approached quality with an open mind, success was possible if the management led the quality improvement process.

The company-wide quality approach places an emphasis on four aspects :-

1. Elements such as controls, job management, adequate processes, performance and integrity criteria and identification of records

2. Competence such as knowledge, skills, experiences, qualifications

3. Soft elements, such as personnel integrity, confidence, organizational culture, motivation, team spirit and quality relationships.

4. Infrastructure (as it enhances or limits functionality)

The quality of the outputs is at risk if any of these aspects is deficient.

QA is not limited to manufacturing, and can be applied to any business or non-business activity, including: design, consulting, banking, insurance, computer software development, retailing, investment, transportation, education, and translation.

It comprises a quality improvement process, which is generic in the sense that it can be applied to any of these activities and it establishes a behavior pattern, which supports the achievement of quality.

This in turn is supported by quality management practices which can include a number of business systems and which are usually specific to the activities of the business unit concerned.

In manufacturing and construction activities, these business practices can be equated to the models for quality assurance defined by the International Standards contained in the ISO 9000 series and the specified Specifications for quality systems.

In the system of Company Quality, the work being carried out was shop floor inspection which did not reveal the major quality problems. This led to quality assurance or total quality control, which has come into being recently.

In Practice

Medical Industry

QA is very important in the medical field because it helps to identify the standards of medical equipments and services. Hospitals and laboratories make use of external agencies in order to ensure standards for equipment such as X-ray machines, Diagnostic Radiology and AERB.

Software Development

Software Quality Assurance consists of a means of monitoring the software engineering processes and methods used to ensure quality. The methods by which this is accomplished are many and varied, and may include ensuring conformance to one or more standards, such as ISO 9000 or a model such as CMMI. In addition, enterprise quality management software is used to correct issues such as: supply chain disaggregation and regulatory compliance which are vital among medical device manufacturers.

Using Contractors and/or Consultants

Consultants and contractors are sometimes employed when introducing new quality practices and methods, particularly where the relevant skills and expertise are not available within the organization or when allocating the available internal resources are not available. Consultants and contractors will often employ Quality Management Systems (QMS), auditing and procedural documentation writing CMMI, Six Sigma, Measurement Systems Analysis (MSA), Quality Function Deployment (QFD), Failure Mode and Effects Analysis (FMEA), and Advance Product Quality Planning (APQP).

References

- "Managing Quality Across the Enterprise: Enterprise Quality Management Solution for Medical Device Companies". Sparta Systems. 2015-02-02.
- Evans, James R. (1994) Introduction to Statistical Process Control Archived October 29, 2013, at the Wayback Machine., Fundamentals of Statistical Process Control, pp 1–13
- Thareja, Mannu; Thareja, Priyavrat (February 2007). "The Quality Brilliance Through Brilliant People". Quality World. 4 (2). Retrieved 2010-01-11.

Sub-Disciplines of Industrial Engineering

Industrial engineering is an umbrella discipline which branches out into unique sub-disciplines which can be studied individually as well as in conjunction with industrial engineering. This all-inclusive chapter elucidates the most significant branches of industrial engineering such as safety engineering, business engineering, enterprise engineering, and engineering management amongst others.

Safety Engineering

International Space Station
Probability of No Impacts From a > 1 cm Ø Debris

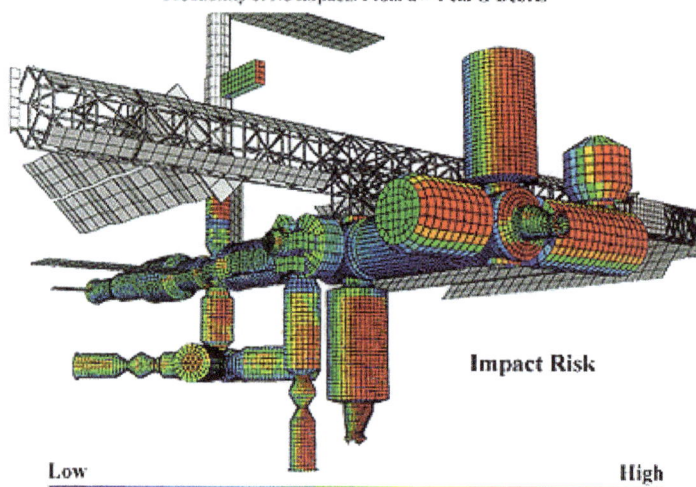

Impact Risk

Low   High

NASA's illustration showing high impact risk areas for the International Space Station

Safety engineering is an engineering discipline which assures that engineered systems provide acceptable levels of safety. It is strongly related to industrial engineering/systems engineering, and the subset system safety engineering. Safety engineering assures that a life-critical system behaves as needed, even when components fail.

Analysis Techniques

Analysis techniques can be split into two categories: qualitative and quantitative methods. Both approaches share the goal of finding causal dependencies between a hazard on system level and failures of individual components. Qualitative approaches focus on the question "What must go wrong, such that a system hazard may occur?", while quantitative methods aim at providing estimations about probabilities, rates and/or severity of consequences.

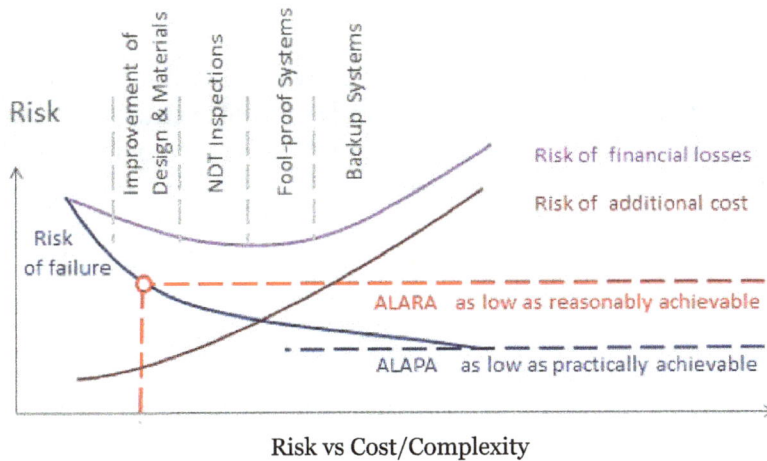

Risk vs Cost/Complexity

The complexity of the technical systems such as Improvements of Design and Materials, Planned Inspections, Fool-proof design, and Backup Redundancy decreases risk and increases the cost. The risk can be decreased to ALARA (as low as reasonably achievable) or ALAPA (as low as practically achievable) levels.

Traditionally, safety analysis techniques rely solely on skill and expertise of the safety engineer. In the last decade model-based approaches have become prominent. In contrast to traditional methods, model-based techniques try to derive relationships between causes and consequences from some sort of model of the system.

Traditional Methods for Safety Analysis

The two most common fault modeling techniques are called failure mode and effects analysis and fault tree analysis. These techniques are just ways of finding problems and of making plans to cope with failures, as in probabilistic risk assessment. One of the earliest complete studies using this technique on a commercial nuclear plant was the WASH-1400 study, also known as the Reactor Safety Study or the Rasmussen Report.

Failure Modes and Effects Analysis

Failure Mode and Effects Analysis (FMEA) is a bottom-up, inductive analytical method which may be performed at either the functional or piece-part level. For functional FMEA, failure modes are identified for each function in a system or equipment item, usually with the help of a functional block diagram. For piece-part FMEA, failure modes are identified for each piece-part component (such as a valve, connector, resistor, or diode). The effects of the failure mode are described, and assigned a probability based on the failure rate and failure mode ratio of the function or component. This quantiazation is difficult for software ---a bug exists or not, and the failure models used for hardware components do not apply. Temperature and age and manufacturing variability affect a resistor; they do not affect software.

Failure modes with identical effects can be combined and summarized in a Failure Mode Effects Summary. When combined with criticality analysis, FMEA is known as Failure Mode, Effects, and Criticality Analysis or FMECA, pronounced "fuh-MEE-kuh".

Fault Tree Analysis

Fault tree analysis (FTA) is a top-down, deductive analytical method. In FTA, initiating primary events such as component failures, human errors, and external events are traced through Boolean logic gates to an undesired top event such as an aircraft crash or nuclear reactor core melt. The intent is to identify ways to make top events less probable, and verify that safety goals have been achieved.

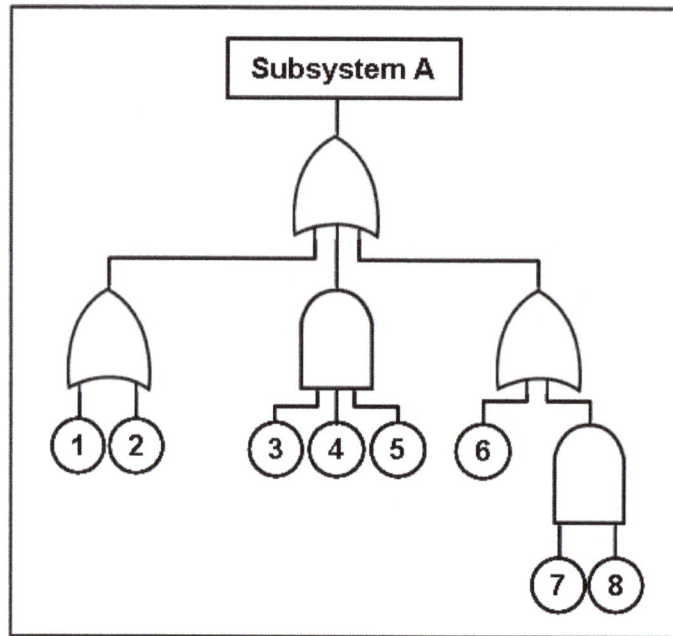

A fault tree diagram

Fault trees are a logical inverse of success trees, and may be obtained by applying de Morgan's theorem to success trees (which are directly related to reliability block diagrams).

FTA may be qualitative or quantitative. When failure and event probabilities are unknown, qualitative fault trees may be analyzed for minimal cut sets. For example, if any minimal cut set contains a single base event, then the top event may be caused by a single failure. Quantitative FTA is used to compute top event probability, and usually requires computer software such as CAFTA from the Electric Power Research Institute or SAPHIRE from the Idaho National Laboratory.

Some industries use both fault trees and event trees. An event tree starts from an undesired initiator (loss of critical supply, component failure etc.) and follows possible further system events through to a series of final consequences. As each new event is considered, a new node on the tree is added with a split of probabilities of taking either branch. The probabilities of a range of "top events" arising from the initial event can then be seen.

Safety Certification

Usually a failure in safety-certified systems is acceptableif, on average, less than one life per 10^9 hours of continuous operation is lost to failure. Most Western nuclear reactors, medical equipment, and commercial aircraft are certifiedto this level. The cost versus loss of lives has been considered appropriate at this level (by FAA for aircraft systems under Federal Aviation Regulations).

Preventing Failure

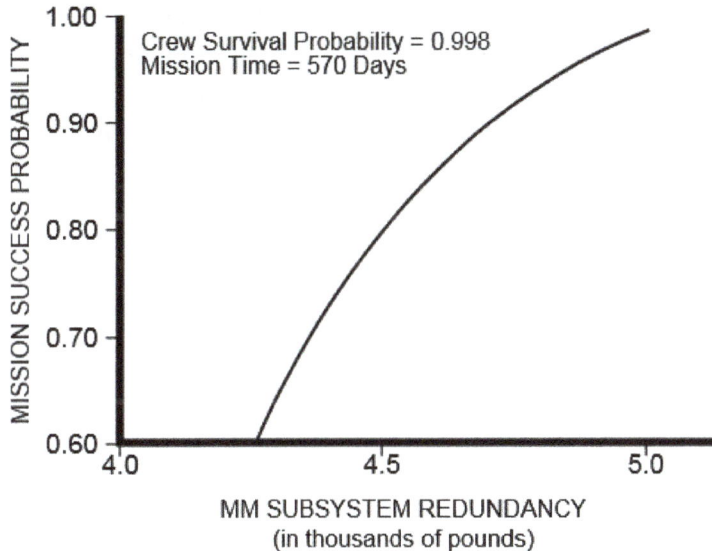

A NASA graph shows the relationship between the survival of a crew of astronauts and the amount of redundant equipment in their spacecraft (the "MM", Mission Module).

Once a failure mode is identified, it can usually be mitigated by adding extra or redundant equipment to the system. For example, nuclear reactors contain dangerous radiation, and nuclear reactions can cause so much heat that no substance might contain them. Therefore, reactors have emergency core cooling systems to keep the temperature down, shielding to contain the radiation, and engineered barriers (usually several, nested, surmounted by a containment building) to prevent accidental leakage. Safety-critical systems are commonly required to permit no single event or component failure to result in a catastrophic failure mode.

Most biological organisms have a certain amount of redundancy: multiple organs, multiple limbs, etc.

For any given failure, a fail-over or redundancy can almost always be designed and incorporated into a system.

There are two categories of techniques to reduce the probability of failure: Fault avoidance techniques increase the reliability of individual items (increased design margin, de-rating, etc.). Fault tolerance techniques increase the reliability of the system as a whole (redundancies, barriers, etc.).

Safety and Reliability

Safety engineering and reliability engineering have much in common, but safety is not reliability. If a medical device fails, it should fail safely; other alternatives will be available to the surgeon. If the engine on a single-engine aircraft fails, there is no backup. Electrical power grids are designed for both safety and reliability; telephone systems are designed for reliability, which becomes a safety issue when emergency (e.g. US "911") calls are placed.

Probabilistic risk assessment has created a close relationship between safety and reliability. Component reliability, generally defined in terms of component failure rate, and external event probability are both used in quantitative safety assessment methods such as FTA. Related probabilistic methods are used to determine system Mean Time Between Failure (MTBF), system availability, or probability of mission success or failure. Reliability analysis has a broader scope than safety analysis, in that non-critical failures are considered. On the other hand, higher failure rates are considered acceptable for non-critical systems.

Safety generally cannot be achieved through component reliability alone. Catastrophic failure probabilities of 10^{-9} per hour correspond to the failure rates of very simple components such as resistors or capacitors. A complex system containing hundreds or thousands of components might be able to achieve a MTBF of 10,000 to 100,000 hours, meaning it would fail at 10^{-4} or 10^{-5} per hour. If a system failure is catastrophic, usually the only practical way to achieve 10^{-9} per hour failure rate is through redundancy.

When adding equipment is impractical (usually because of expense), then the least expensive form of design is often "inherently fail-safe". That is, change the system design so its failure modes are not catastrophic. Inherent fail-safes are common in medical equipment, traffic and railway signals, communications equipment, and safety equipment.

The typical approach is to arrange the system so that ordinary single failures cause the mechanism to shut down in a safe way (for nuclear power plants, this is termed a passively safe design, although more than ordinary failures are covered). Alternately, if the system contains a hazard source such as a battery or rotor, then it may be possible to remove the hazard from the system so that its failure modes cannot be catastrophic. The U.S. Department of Defense Standard Practice for System Safety (MIL–STD–882) places the highest priority on elimination of hazards through design selection.

One of the most common fail-safe systems is the overflow tube in baths and kitchen sinks. If the valve sticks open, rather than causing an overflow and damage, the tank spills into an overflow. Another common example is that in an elevator the cable supporting the car keeps spring-loaded brakes open. If the cable breaks, the brakes grab rails, and the elevator cabin does not fall.

Some systems can never be made fail safe, as continuous availability is needed. For example, loss of engine thrust in flight is dangerous. Redundancy, fault tolerance, or recovery procedures are used for these situations (e.g. multiple independent controlled and fuel fed engines). This also makes the system less sensitive for the reliability prediction errors or quality induced uncertainty for the separate items. On the other hand, failure detection & correction and avoidance of common cause failures becomes here increasingly important to ensure system level reliability.

Associations

- Institute of Industrial Engineers

Business Engineering

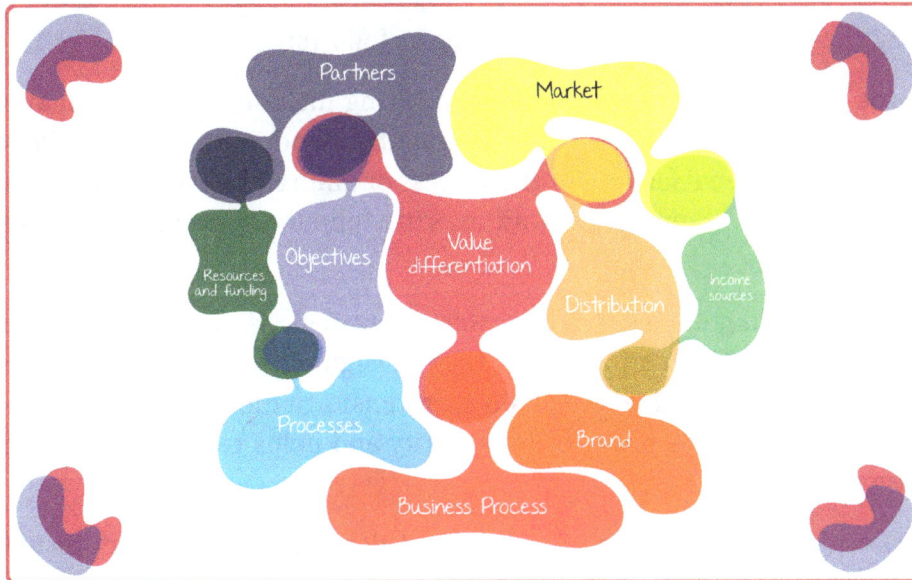

The Business life model, an example of business engineering, is a tool created to help entrepreneurs, business people and academics build stronger business models.

Business Engineering (BE) refers to the development and implementation of business solutions, from business model to business processes and organizational structure to information systems and information technology (cf.).

Business engineering focuses on developing innovative business solutions that take a sociotechnical systems (STS) approach. The STS approach addresses an enterprise as a total system in the same manner as an airplane system or an industrial facility, since they possess a similar level of complexity.

Business engineering combines knowledge in the fields of business administration, Industrial Engineering, as well as information technology and connects it to all aspects of transformation, from means of presentation to process models to cultural and political considerations (cf. Baumöl/ Jung).

Overview

Business engineering focuses on challenges arising from the transformation of the industrial society into an information society (cf. Winter), that is the digitization of enterprises, economy, administration and society. Through the ongoing consumerization digital services for individuals have also become a crucial part of research (cf. Hess/Legner). Because of the major importance of information technology, business engineering is often held to be a subfield of Business Informatics, although it is also sometimes regarded as a form of Organization Development for its emphasis on Change Management.

Engineering Management is a very close discipline which overlaps significantly with Business

Engineering; the main differences are that Industrial Engineering focuses primarily on the goods sector (less on services), on technical systems and the interface between those systems as well as the users from a production point of view.

Österle/Blessing characterize business engineering as follows:

- Beside the technical design, business engineering includes the political and cultural dimensions of a new business solution. The political and cultural dimensions and change management are crucial factors for the success or failure of a transformation (cf. Baumöl). Therefore, business engineering is an interdisciplinary approach. It divides the design levels of a company.

- Business engineering distinguishes between a strategic, an organizational and a technological design level. Contemplating different design objects on different levels enables a focused view of the individual dimensions of transformation (, S. 191). Segmenting task at hand into different levels provides for security and helps reducing the complexity of the transformation process.

- Business engineering ensures a holistic view of all dimensions. It supports not only the design of new business models, business processes and information systems, but also their implementation. Therefore, it contemplates all dimensions (resources and processes involved) of the transformation.

- Business engineering refers to the method and model-based design theory for companies in the information age (, S. 7). Business transformations along with their technical and socio-economic aspects are far too important and complex to be realized without applying methods and models. Methods and models not only provide for transparency during the process of transformation, they also specify the division of labor, create a foundation for communication and enable the documentation of the company's systematic reorientation. The division of labor and application of engineering principles differentiate the "construction" in accordance with business engineering from individualistic "creation" (cf., S. 88).

- Business engineering focuses on the consumer from a business perspective. As of now, this also holds true for the deep penetration of all spheres of private life with information technology (consumerization), which is equally being treated from a business and not an individual point of view.

At the Business Engineering Forum in St. Gallen, experts from science and business annually discuss new developments within the discipline.

Approaches to Business Engineering

ARIS (Architecture of Integrated Information Systems)

The ARIS concept (cf. Scheer) distinguishes between views and levels of description. The views are the organizational, data, performance, functional and control view of processes. For every description view there are three distinct levels of description. These are the technical concept, the data processing concept and the implementation level. Through this framework, specific methods can be used to describe individual elements of the model. Those methods help depict and improve business processes, from fundamental business questions to the implementation on the IT level.

Business Engineering St.Gallen

The St. Gallen approach to Business Engineering comprises fundamentals and methods for different kinds of transformation projects. It distinguishes between the design level strategies, organization and information systems within the transformation process and, thus, reduces the complexity of the transformation.

- Strategy: Capabilities (incl. brand) and resources, business segments (products, services, customer segments), customer access, competitive position, ecosystem, and revenue and cost structures

- Organization: company organization structure and operational structure, focus on business processes with accompanying process performances, procedures, tasks and business objects

- Information system: System of applications and technical services, software and data components as well as IT infrastructure components. According to the relevance of IT in the respective organization, this level is further divided into the subcategories of "alignment", "software and data" and "IT infrastructure" (cf. Winter/Fischer).

The St. Gallen approach to Business Engineering has already been applied in more than 1.000 consultancy projects (S&T 2008). Moreover, it receives support from renowned software tools such as the ARIS toolset (Scheer 2001, S. 18), ADOben (Aier et al. 2008) or Semtalk.

Semantic Object Model (SOM)

The Semantic Object Model (SOM) (cf. Ferstl/Sinz) provides a framework for the conceptual modeling of system types. The modeling concept differentiates three levels of such systems: The strategic business plan; the operational business process models which focus on process types; and the specifications for implementing application systems. These levels can also be regarded as constituting perspectives on the system. They correspond to the external, internal, and resource perspectives. Subsequently, the levels of the model undergo a further conceptual division according to the organizational architecture, the process model, the principles of coordination or the object system.

The Open Group Architecture Framework (TOGAF)

The Open Group Architecture Framework (TOGAF) (cf. Weinberger) presents a structure for organizational architectures which offers a holistic approach to designing, planning, implementing and maintaining information architectures and, thus, covers an important section, albeit not the entire scope, of BE. When applying TOGAF, the enterprise architecture is usually modeled in the three domains business architecture, information systems architecture (consisting of application architecture and data architecture), and technology architecture.

Integral Design of Organizational Structures and Information Systems

Van Meel and Sol (1996) define business engineering as an „integral design of organizational architectures and information systems". Dynamic modeling as well as simulation constitute the key components of their approach.

The3rdgear Business Engineering Model

With a specific focus on the demands of small business, The3rdgears' Business Engineering system has proven to be effective and relatively easy to implement for small to medium-sized businesses. Unique in that it involves the use of a specialised customisable software platform that integrates key business structures and operational performance processes. While The3rdgear system encapsulates the key areas of Strategy, Business Process, Technology and People, it has been proven to provide unusually consistent, predictable outcomes for small businesses due to the SME specific design of its improvement processes and software applications. This, combined with its ease of use for business owners and managers has made The3rdgear model popular within the SME marktet place.

Education

In Belgium

In Belgium, there are numerous institutions offering bachelor and master programs in business engineering, among them the Solvay Brussels School of Economics and Management (Université Libre de Bruxelles and Vrije Universiteit Brussel), the ICHEC Brussels Management School, the Louvain School of Management (Université catholique de Louvain, Université de Namur), the HEC Management School - University of Liège, the KULeuven, the HUBrussel (Hogeschool-Universiteit Brussel), the Ghent University, the University of Antwerp and the University of Hasselt. The programs contain both management methodology, business administration, economics, mathematics, and technical elements and computer science.

In Finland

In Finland, the University of Oulu offers a diploma program in Business Engineering for which a master's degree in Engineering is required.

In Mexico

In Mexico, the Instituto Tecnológico Autónomo de México (ITAM) offers a professional Business Engineering program that contains business administration, mathematics, marketing, economics, finances, and IT and computer science. Students focus in entrepreneurship and business consulting.

In Germany

In Germany, the Steinbeis University offers a master's program in Business Engineering. The contents of the degree course held in English include, similar to Belgium, management methodologies as well as fundamentals of business administration and economics. As in Finland, an academic degree in Engineering is required in order to be eligible for admission.

In Switzerland

In Switzerland, professionals can receive the degree of Executive MBA in Business Engineering at the University of St. Gallen or at the PHW Hochschule Wirtschaft Bern. The program at the University of St. Gallen lasts 18 months and includes two modules in the USA (Santa Clara University) and China (Shanghai Jiao Tong University) where students learn about the local dynamics, culture and differences in management methods compared to Europe. The program's main focus lies on providing practical knowledge in the areas of business strategy/business model, business processes/organizational structure, information systems/information technology, and the political and cultural dimensions of economic activity. The systematic exchange of experience among the students complements the educational concept. At the end of the program, the newly received Business Engineers are qualified to actively and holistically transform their businesses and take over executive positions within their organizations. To be eligible for admission, professionals must have an academic degree and about 5 years of work experience. The program at the PHW Hochschule Wirtschaft Bern can be completed in 4 semesters and focuses on strategies of internationalization, technology management, business models and value added concepts, information management and systems, value-centered corporate governance and turnaround management.

In Chile

The University of Chile offers a master's program in Business Engineering (M. Sc.) since 2003. As in Germany and Belgium the program combines management theory, business administration, finance, economics, science and technology. The program consists of a theoretical as well as a practical part during which the students develop a project for an organization in the private or public sector. The program is designed to be completed in four semesters. Hundreds of projects have been developed since 2003, producing a validation of the proposed approach to design and a generation of knowledge that has been formalized and applied in new projects. Graduates are awarded with the academic degree "Master of Science in Business Engineering" (MBE) after taking the courses of the program and successfully completing their projects.

In Peru

The University of the Pacific offers an academic degree in business engineering since 2008. The students deepen their knowledge in the three areas of Process Engineering, Project Engineering and Information Technology Engineering while additionally being able to choose from a wide variety of subjects to complement their studies, e.g. mathematics, physics, informatics, economics, philosophy, sociology etc.

In El Salvador

The Escuela Superior de Economia y Negocios (ESEN) initiated a business engineering program for professionals in 2009 which focuses on the systems perspective, the modeling and analysis of complex relationships between resources, employees and information, as well as the integration of engineering expertise with a quantitative and qualitative business understanding. The program combines elements of different disciplines, e.g. business administration, economics, management, innovation and information technology. The career has a duration of 5 years with the possibility that in the third cycle of the 4th year, the student can study in Germany, Chile, United States and Mexico as long as your CUM is up 8.0.

In The Philippines

The Ateneo de Naga University offers a bachelor program in business administration (B. Sc.) with the possibility to specialize in business engineering. The four-year program aims at training young technology-based founders and students who want to pursue a career in technology-related industries. Practice-oriented projects encourage the creativity and innovative energy needed to conceptualize and implement products and ideas while advancing the socio-economic development of the Bicol region at the same time. Moreover, the university maintains close relations to entrepreneurs and businessmen in order to enable the students to learn from "best practices".

Enterprise Engineering

Enterprise engineering is defined as the body of knowledge, principles, and practices to design an enterprise. An enterprise is a complex, socio-technical system that comprises interdependent resources of people, information, and technology that must interact with each other and their environment in support of a common mission. Enterprise engineering is a subdiscipline of industrial engineering / systems engineering. The discipline examines each aspect of the enterprise, including business processes, information flows, and organizational structure. Enterprise engineering may focus on the design of the enterprise as a whole, or on the design and integration of certain business components.

Overview

In theory and practice more types of enterprise engineering have emerged. In the field of engineering, a more general form of enterprise engineering has emerged. Encompassing "the application of knowledge, principles, and disciplines related to the analysis, design, implementation and operation of all elements associated with an enterprise. In essence this is an interdisciplinary field which combines systems engineering and strategic management as it seeks to engineer the entire enterprise in terms of the products, processes and business operations,". this field is related to engineering management, operations management, service management and systems engineering.

In the context of software development, a specific field of enterprise engineering has also appeared that deals with the modelling and integration of various organizational and technical parts of business processes and functions. In the context of information systems development, this has become an area of activity for the organization of systems analysis, and an extension to the existing scope of Information Modelling. It can also be viewed as an extension and generalization of the systems analysis and systems design phases of the software development process. Here, enterprise modelling can form part of the early, middle and late information system development life cycle. Explicit representation of the organizational and technical system infrastructure is being developed in order to understand the orderly transformations of existing work practices. This discipline is also known as Enterprise architecture, or along with Enterprise ontology, defined as being one of the two major sub-fields of Enterprise architecture.

In a 2013 article of enterprise engineering has been published. It conveys the ideas of the CIAO! Network (www.ciaonetwork.org). The discipline of enterprise engineering as defined in this article comprises all of the above-mentioned fields. Three major objectives are proposed: intellectual manageability, organizational concinnity, and social devotion.

Enterprise Engineering Methods

Enterprise engineering involves formal methodologies, methods and techniques which are designed, tested and used extensively in order to offer organizations reusable business process solutions:

- Computer Integrated Manufacturing Open Systems Architecture (CIMOSA) methodology

- Integrated DEFinition (IDEF) methodology

- Petri Nets

- Unified Modeling Language (UML) or Unified Enterprise Modeling Language (UEML)

- Enterprise Function Diagrams (EFD)

These methodologies, techniques and methods are all more or less suited to modeling an enterprise and its underlying processes.

Computer Integrated Manufacturing Open Systems Architecture

CIMOSA provides templates and interconnected modeling constructs to encode business, people and information technology (IT) aspects of enterprise requirements. This is done from multiple perspectives: Information view, Function view, Resource view and Organization view. These constructs can further be used to structure and facilitate the design and implementation of detailed IT systems.

The division into different views makes it a clarifying reference for enterprise and software engineers. It shows information needs for different enterprise functions such as activities, processes and operations alongside their corresponding resources. In this way it can easily be determined which IT system will fulfill the information needs of a particular activity and its associated processes.

IDEF

IDEF, first developed as a modeling language to model manufacturing systems, has been used by the U.S. Airforce since 1981 and originally offered four different notations to model an enterprise from a certain viewpoint. These were IDEF0, IDEF1, IDEF2 and IDEF3 for functional, data, dynamic and process analysis respectively. Over the past decades a number of tools and techniques for the integration of these different notations have been developed incrementally.

IDEF shows how a business process flows through a variety of decomposed business functions with corresponding information inputs, outputs and actors. Like CIMOSA, it also uses different enterprise views. Moreover, IDEF can be easily transformed into UML-diagrams for the further development of IT systems. These positive characteristics make it a powerful method for the development of Functional Software Architectures.

Petri Nets

Petri Nets are established tools used to model manufacturing systems. They are highly expressive and provide good formalisms for the modeling of concurrent systems. The most advantageous properties are the ability to create simple representation of states, concurrent system transitions and capabilities thereby allowing modelling of the duration of transitions. As a result, Petri Nets can be used to model certain business processes with corresponding state and transitions or activities therein as well as outputs. Moreover, Petri Nets can be used to model different software systems and transitions between these systems. In this way programmers can use it as a schematic coding reference.

In recent years research has shown that Petri Nets can contribute to the development of business process integration. One of these is the "Model Blue" methodology developed by IBM's Chinese Research Laboratory. Model Blue outlines the importance of model driven business integration as an emerging approach to building integrated software platforms. The correspondence between their Model Blue business view and an equivalent Petri Net is also shown, which indicates that their research has closed the gap between business and IT. However, instead of Petri Nets the researchers instead use their own Model Blue IT view, which can be derived from their business view through a transformation engine.

Unified Modeling Language (UML)

Unified Modeling Language (UML) is a broadly accepted modeling language for the development of software systems and applications. Many within the Object-oriented analysis and design community also use UML for enterprise modeling purposes. Here, emphasis is placed on the usage of enterprise objects or business objects from which complex enterprise systems are made. A collection of these objects and corresponding interactions between them can represent a complex business system or process. While Petri Nets focus on the interaction and states of objects, UML focuses more on the business objects themselves. Sometimes these are called "enterprise building blocks" and includes resources, processes, goals, rules and metamodels. Despite the fact that UML can be used to model an integrated software system, it has been argued that the reality of business can be modeled with a software modeling language. In response, the object oriented community makes business extensions for UML and adapts the language accordingly. Extended Enterprise Modeling Language (EEML) is derived from UML and is proposed as a business modeling language. The question remains as to whether this business transformation is the correct method to use, as it was earlier said that UML in combination with other "pure' business methods may be a better alternative.

Enterprise Function Diagrams

EFD is a used as a modeling technique for the representation of enterprise functions and corresponding interactions. Different business processes can be modeled in these representations through the use of "function modules" and triggers. A starting business process delivers different inputs to different functions. A process flowing through all the functions and sub-functions creates multiple outputs. Enterprise Function Diagrams thereby provide an easy-to-use and detailed representation about a business process and its corresponding functions, inputs, outputs and triggers. In this way EFD has many similarities with IDEFo diagrams, which also represent business processes in a hierarchical fashion as a combination of functions and triggers. The two differ in that an EFD places the business functions in an organization hierarchical perspective, which outlines the downstream of certain processes in the organization. On the other hand, IDEFo diagrams show the responsibilities of certain business functions through the use of arrows. Furthermore, IDEFo provides a clear representation of inputs and outputs for every (sub) functionw.

EFD may be used as a business front-end to a software modeling language like UML and its major similarities to IDEF as a modeling tool indicate that this is indeed possible. However, further research is needed to improve EFD techniques in such a way that formal mappings to UML can be made. Research on the complementary use of IDEF and UML has contributed to the acceptance of IDEF as business-front end and therefore a similar study should be carried out with EFD and UML.

Associations

- INFORMS
- Institute of Industrial Engineers

Engineering Management

Engineering management is a specialized form of management that is concerned with the application of engineering principles to business practice. Engineering management is a career that brings together the technological problem-solving savvy of engineering and the organizational, administrative, and planning abilities of management in order to oversee complex enterprises from conception to completion. A Master of Science in Engineering Management (MSEM, or MS in Engineering Management) is sometimes compared to a Master of Business Administration (MBA) for professionals seeking a graduate degree as a qualifying credential for a career in engineering management.

Ron Diftler (left), NASA Robonaut manager assisting in a Robonaut familiarization training session in the Space Environment Simulation Laboratory at NASA's Johnson Space Center.

Engineering management is considered to be a subdiscipline of industrial engineering/systems engineering. Successful engineering managers typically require training and experience in business and engineering. Technically inept managers tend to be deprived of support by their technical team, and non-commercial managers tend to lack commercial acumen to deliver in a market economy. Largely, engineering managers manage engineers who are driven by non-entrepreneurial thinking, and thus require the necessary people skills to coach, mentor and motivate technical professionals. Engineering professionals joining manufacturing companies sometimes become engineering managers by default after a period of time. They are required to learn how to manage once they are on the job, though this is usually an ineffective way to develop managerial abilities.

History

Stevens Institute of Technology is believed to have the oldest Engineering management department, established as the School of Business Engineering in 1908. This was later called the Bachelor of Engineering in Engineering Management (BEEM) program and moved into the School of Systems and Enterprises. Drexel University established the first graduate engineering management degree in the U.S., which was first offered in 1959. In 1967 the first university department titled "Engineering Management" was founded at the Missouri University of Science and Technology (Missouri S&T, formerly the University of Missouri-Rolla, formerly Missouri School of Mines).

Outside the USA, in Germany the first department concentrating on Engineering Management was established 1927 in Berlin. In Turkey the Istanbul Technical University has a Management Engineering Department established in 1982, offering a number of graduate and undergraduate programs in Management Engineering. In UK the University of Warwick has a specialised department WMG (previously known as Warwick Manufacturing Group) established in 1980, which offers a graduate programme in MSc Engineering Business Management.

More recently in the United Kingdom, Teesside University's School of Science and Engineering introduced an MSc Engineering Management alongside its engineering-focused MSc Project Management.

Michigan Technological University began an Engineering Management program in the School of Business & Economics in the Fall of 2012.

In Canada, Memorial University of Newfoundland has started a complete master's degree Program in Engineering Management Master of Engineering Management.

In Denmark, the Technical University of Denmark offers a MSc program in Engineering Management (in English). MSc in Engineering Management.

In Pakistan, NED University of Engineering & Technology, Karachi has been running a Master of Engineering in Engineering Management program since 2005. A variant of this program is within Quality Management. COMSATS (CIIT) offers a MSc Project Management program to Local and Overseas Pakistanis as an on-campus/off-campus student.

Areas

Operations Research and Supply Chain Management

Operations research deals with quantitative models of complex operations and uses these models to support decision-making in any sector of industry or public services. Supply chain management is the process of planning, implementing and managing the flow of goods, services and related information from the point of origin to the point of consumption.

Information Technologies

The "information technologies" theme focuses on how technology is designed and managed to support effective decision-making. Topics deal with technical applications in software design and development, data mining and telecommunication as well as the organizational and social issues associated with the use of information technologies.

Decision Engineering

Decision engineering seeks to use engineering principles in the creation of a decision, which it views as an engineering artifact in its own right. From this point of view, the creation of a decision includes agreeing to objectives, developing a detailed specification, and then creating a decision model, which captures the key cause-and-effect elements of the decision environment (a systems thinking approach) with a focus on the particular decision, instead of the entire system (which can be otherwise intractable). Like other engineered artifacts, a decision model can be subject to Quality assurance review, and-since it is documented-is amenable to Process improvement over time. Decision engineering models draw from the information technologies described above for data supporting the decision, but are distinguished from IT in that they model the decision, not just the data supporting it.

Management of Technology

The Management of Technology (MOT) theme builds on the foundation of management topics in accounting, finance, economics, organizational behavior and organizational design. Courses in this theme deal with operational and organizational issues related to managing innovation and technological change.

Education

Engineering Management programs typically include instruction in accounting, economics, finance, project management, systems engineering, mathematical modeling and optimization, management information systems, quality control & six sigma, operations research, human resources management, industrial psychology, safety and health.

There are many options for entering into engineering management, albeit that the foundation requirement is an engineering degree (or other computer science, mathematics or science degree) and a business degree.

Undergraduate Degrees

Although most engineering management programs are geared for graduate studies, there are a few elite institutions that teach EM at the undergraduate level. Some of the ones that are accredited and/or recognized by ASEM include: West Point (United States Military Academy), Norwich University, New York Institute of Technology, Stevens Institute of Technology, Illinois Institute of Technology, Rensselaer Polytechnic Institute, George Washington University, Arizona State University, Missouri University of Science and Technology. Graduates of these programs regularly command nearly $65,000 their first year out of school.

Outside the USA, Istanbul Technical University Management Engineering Department offers an elite undergraduate degree in Management Engineering, attracting top students. The University of Waterloo offers a 4-year undergraduate degree (5 years including co-op education) in the field of Management Engineering. This is the first program of its kind in Canada. In Peru, Pacifico University offers a 5-year undergraduate degree in this field, the first program in this country.

Graduate Degrees

Many universities offer Master of Engineering Management degrees. Missouri S&T is credited with awarding the first Ph.D. in Engineering Management in 1984. The National Institute of Industrial Engineering based in Mumbai has been awarding degrees in the field of Post Graduate Diploma in Industrial Engineering since 1973 and the Fellowship (Doctoral) degrees have been awarded since 2008. Massachusetts Institute of Technology offers a Master in System Design and Management, which is a member of the Consortium of Engineering Management.

Students in the University of Kansas' Engineering Management Program are practicing professionals employed by over 100 businesses, manufacturing, government or consulting firms. There are over 200 actively enrolled students in the program and approximately 500 alumni.

Istanbul Technical University Management Engineering Department offers a graduate degree, and a Ph.D. degree in Management Engineering.

According to the American Society for Engineering Education (ASEE) PRISM Magazine (March 2008) the largest Master's of Engineering Management (MEM) programs (in terms of degrees awarded for 2005 -2006) are shown in the following chart.

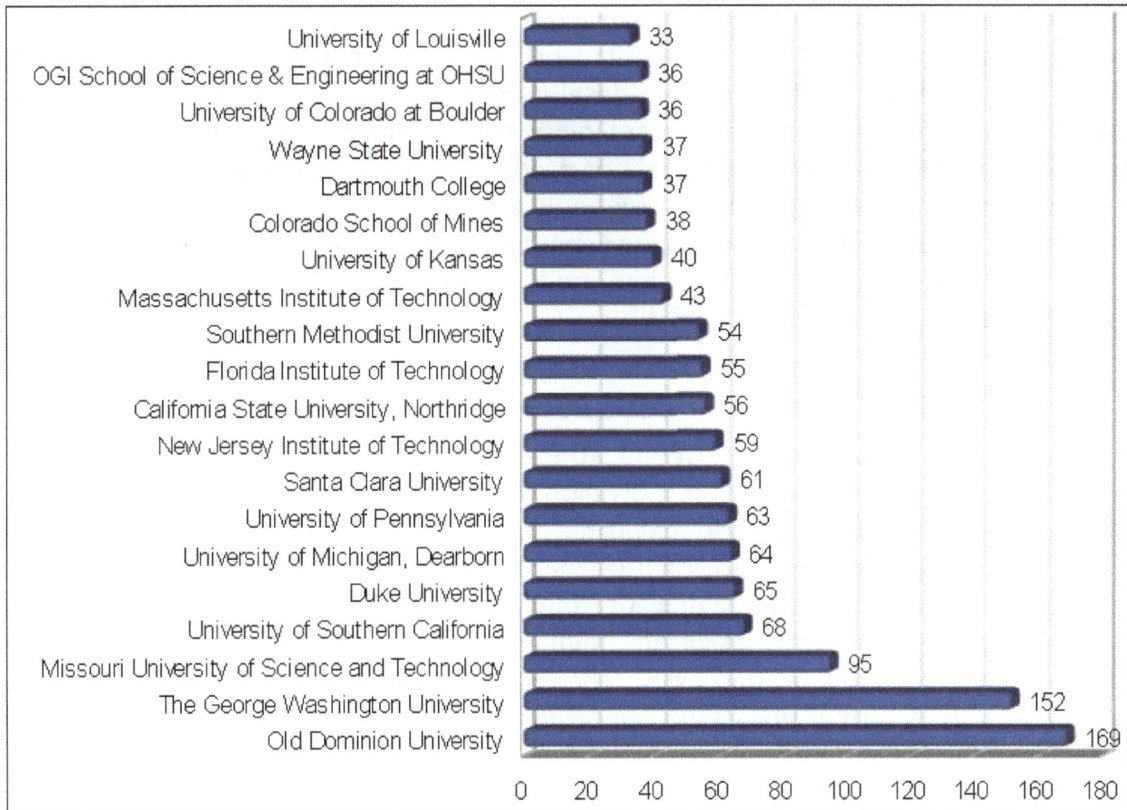

University	Value
University of Louisville	33
OGI School of Science & Engineering at OHSU	36
University of Colorado at Boulder	36
Wayne State University	37
Dartmouth College	37
Colorado School of Mines	38
University of Kansas	40
Massachusetts Institute of Technology	43
Southern Methodist University	54
Florida Institute of Technology	55
California State University, Northridge	56
New Jersey Institute of Technology	59
Santa Clara University	61
University of Pennsylvania	63
University of Michigan, Dearborn	64
Duke University	65
University of Southern California	68
Missouri University of Science and Technology	95
The George Washington University	152
Old Dominion University	169

Engineering Management Consulting

As engineering firms are usually small partnerships, they cannot afford in-house management, therefore giving rise to the need for engineering management consultancy. It involves providing management consulting advice that is specific to engineering. Indifferent from the traditional focus of the, A T Kearney, Boston Consulting Group and McKinsey, science and engineering requires a particularly holistic approach involving art and science. There are many branches of engineering management consultancy (commerce), including law, accounting, human resources, marketing, politics, economics, finance, public affairs, and communication. Commonly, engineering management consultants are also used when firms require special technical knowledge, though many prefer to use engineering educational consultants for such a task, to upgrade organizational knowledge and in able to keep the intellectual property confidential. Though many firms opt to use traditional management consulting firms, many lack the know-how to tailor the traditional theories to accommodate professional engineers and other technical workers.

Engineering management consulting is concerned with the development, improvement, implementation and evaluation of integrated systems of people, money, knowledge, information, equipment, energy, materials and/or processes. Consultants strive to improve upon existing processes, products or systems. Engineering management consulting draws upon the principles and methods of engineering analysis and synthesis, as well as the mathematical, physical and social sciences together with the principles and methods of engineering design to specify, predict, and evaluate the results to be obtained from such systems or processes. Engineering management consulting puts a focus on the social impact of the product, process or system that is being analyzed.

Examples of where engineering management consulting might be used include designing new product development process, or an assembly workstation, strategizing for various operational logistics, consulting as an efficiency expert, developing a new financial algorithm or loan system for a bank, streamlining operation and emergency room location or usage in a hospital, planning complex distribution schemes for materials or products (referred to as Supply Chain Management), and shortening lines (or queues) at a bank, hospital, or a theme park. Management engineering consultants typically use computer simulation (especially discrete event simulation), along with extensive mathematical tools and modeling and computational methods for system analysis, evaluation, and optimization.

Professional Organizations

There are a number of societies and organizations dedicated to the field of engineering management. One of the largest societies is a division of IEEE, the Engineering Management Society, which regularly publishes a trade magazine. Another prominent professional organization in the field is the American Society for Engineering Management (ASEM), which was founded in 1979 by a group of 20 engineering managers from industry. ASEM currently certifies engineering managers (two levels) via the Associate Engineering Manager (AEM) or Professional Engineering Manager (PEM) certification exam. The Master of Engineering Management Programs Consortium is a newly formed consortium of prominent universities intended to raise the value and visibility of the MEM degree. Also, engineering management university programs have the possibility of being accredited by ABET, ATMAE, or ASEM. In Canada, the Canadian Society for Engineering Management (CSEM) is a constituent society of the Engineering Institute of Canada (EIC), Canada's oldest learned engineering society.

Associations

- Society of engineering and management systems

- Institute of Industrial and Systems Engineers

- INFORMS

References

- Kletz, Trevor (1984). Cheaper, safer plants, or wealth and safety at work: notes on inherently safer and simpler plants. I.Chem.E. ISBN 0-85295-167-1.

- Lutz, Robyn R. (2000). Software Engineering for Safety: A Roadmap (PDF). The Future of Software Engineering. ACM Press. ISBN 1-58113-253-0. Retrieved 31 August 2006.

- Leveson, Nancy (2011). Engineering a Safer World - Systems Thinking Applied To Safety. Engineering Systems. The MIT Press. ISBN 978-0-262-01662-9. Retrieved 3 July 2012.

- Vernadat, F.B. (1996) Enterprise Modeling and Integration: Principles and Applications. Chapman & Hall, London, ISBN 0-412-60550-3.

- Bornschlegl, Susanne (2012). Ready for SIL 4: Modular Computers for Safety-Critical Mobile Applications (pdf). MEN Mikro Elektronik. Retrieved 2015-09-21.

- De Vries, Marne, Aurona Gerber, and Alta van der Merwe. In: Aveiro D., Tribolet J., Gouveia D. (eds) "The Nature of the Enterprise Engineering Discipline." Advances in Enterprise Engineering VIII. Springer International Publishing, 2014. p. 1-15.

- Ulrike Baumöl, Reinhard Jung: Rekursive Transformation: Entwicklung der Business Engineering-Landkarte. In: Walter Brenner, Thomas Hess (Hrsg.): Wirtschaftsinformatik in Wissenschaft und Praxis – Festschrift für Hubert Österle. Business Engineering. Springer, Berlin 2014.

- US DOD (10 February 2000). Standard Practice for System Safety (PDF). Washington, DC: US DOD. MIL-STD-882D. Retrieved 7 September 2013.

- Jan Dietz, Jan Hoogervorst et al. (2013). The Discipline of Enterprise Engineering". International Journal of Organisational Design and Engineering, Vol. 3, No. 1, 2013, pp 86-114.

Industrial Ecology: A Comprehensive Study

Industrial ecology has emerged as an important area of study in the recent decade. This field also focuses on effective ways to reduce industrial waste. The following chapter will present to the reader all the significant aspects of industrial ecology and also elaborate the history of this field for a better understanding.

Industrial Ecology

Industrial ecology (IE) is the study of material and energy flows through industrial systems. The global industrial economy can be modelled as a network of industrial processes that extract resources from the Earth and transform those resources into commodities which can be bought and sold to meet the needs of humanity. Industrial ecology seeks to quantify the material flows and document the industrial processes that make modern society function. Industrial ecologists are often concerned with the impacts that industrial activities have on the environment, with use of the planet's supply of natural resources, and with problems of waste disposal. Industrial ecology is a young but growing multidisciplinary field of research which combines aspects of engineering, economics, sociology, toxicology and the natural sciences.

Industrial ecology has been defined as a "systems-based, multidisciplinary discourse that seeks to understand emergent behaviour of complex integrated human/natural systems". The field approaches issues of sustainability by examining problems from multiple perspectives, usually involving aspects of sociology, the environment, economy and technology. The name comes from the idea that the analogy of natural systems should be used as an aid in understanding how to design sustainable industrial systems.

Overview

Industrial ecology is concerned with the shifting of industrial process from linear (open loop) systems, in which resource and capital investments move through the system to become waste, to a closed loop system where wastes can become inputs for new processes.

Much of the research focuses on the following areas:

- material and energy flow studies ("industrial metabolism")
- dematerialization and decarbonization
- technological change and the environment
- life-cycle planning, design and assessment

- design for the environment ("eco-design")

- extended producer responsibility ("product stewardship")

- eco-industrial parks ("industrial symbiosis")

- product-oriented environmental policy

- eco-efficiency

Industrial ecology seeks to understand the way in which industrial systems (for example a factory, an ecoregion, or national or global economy) interact with the biosphere. Natural ecosystems provide a metaphor for understanding how different parts of industrial systems interact with one another, in an "ecosystem" based on resources and infrastructural capital rather than on natural capital. It seeks to exploit the idea that natural systems do not have waste in them to inspire sustainable design.

Example of Industrial Symbiosis. Waste steam from a waste incinerator (right) is piped to an ethanol plant (left) where it is used as in input to their production process.

Along with more general energy conservation and material conservation goals, and redefining commodity markets and product stewardship relations strictly as a service economy, industrial ecology is one of the four objectives of Natural Capitalism. This strategy discourages forms of amoral purchasing arising from ignorance of what goes on at a distance and implies a political economy that values natural capital highly and relies on more instructional capital to design and maintain each unique industrial ecology.

History

Industrial ecology was popularized in 1989 in a *Scientific American* article by Robert Frosch and Nicholas E. Gallopoulos. Frosch and Gallopoulos' vision was "why would not our industrial system behave like an ecosystem, where the wastes of a species may be resource to another species? Why would not the outputs of an industry be the inputs of another, thus reducing use of raw materials, pollution, and saving on waste treatment?" A notable example resides in a Danish industrial park in the city of Kalundborg. Here several linkages of byproducts and waste heat can be found between numerous entities such as a large power plant, an oil refinery, a pharmaceutical plant, a plasterboard factory, an enzyme manufacturer, a waste company and the city itself.

View of Kalundborg Eco-industrial Park

The scientific field Industrial Ecology has grown quickly in recent years. The Journal of Industrial Ecology (since 1997), the International Society for Industrial Ecology (since 2001), and the journal Progress in Industrial Ecology (since 2004) give Industrial Ecology a strong and dynamic position in the international scientific community. Industrial Ecology principles are also emerging in various policy realms such as the concept of the Circular Economy that is being promoted in China. Although the definition of the Circular Economy has yet to be formalized, generally the focus is on strategies such as creating a circular flow of materials, and cascading energy flows. An example of this would be using waste heat from one process to run another process that requires a lower temperature. The hope is that strategy such as this will create a more efficient economy with fewer pollutants and other unwanted by-products.

Principles

Biosphere	Technosphere
• Environment	• Market
• Organism	• Company
• Natural Product	• Industrial Product
• Natural Selection	• Competition
• Ecosystem	• Eco-Industrial Park
• Ecological Niche	• Market Niche
• Anabolism / Catabolism	• Manufacturing / Waste Management
• Mutation and Selection	• Design for Environment
• Succession	• Economic Growth
• Adaptation	• Innovation
• Food Web	• Product Life Cycle

One of the central principles of Industrial Ecology is the view that societal and technological systems are bounded within the biosphere, and do not exist outside of it. Ecology is used as a *metaphor* due to the observation that natural systems reuse materials and have a largely closed loop cycling of nutrients. Industrial Ecology approaches problems with the hypothesis that by using similar principles as *natural systems, industrial systems* can be improved to reduce their impact on the natural environment as well.

The Kalundborg industrial park is located in Denmark. This industrial park is special because companies reuse each other's waste (which then becomes by-products). For example, the Energy E2 Asnæs Power Station produces gypsum as a by-product of the electricity generation process; this gypsum becomes a resource for the BPB Gyproc A/S which produces plasterboards. This is one example of a system inspired by the biosphere-technosphere metaphor: in ecosystems, the waste from one organism is used as inputs to other organisms; in industrial systems, waste from a company is used as a resource by others.

Apart from the direct benefit of incorporating waste into the loop, the use of an eco-industrial park can be a means of making renewable energy generating plants, like Solar PV, more economical and environmentally friendly. In essence, this assists the growth of the renewable energy industry and the environmental benefits that come with replacing fossil-fuels.

IE examines societal issues and their relationship with both technical systems and the environment. Through this *holistic view* , IE recognizes that solving problems must involve understanding the connections that exist between these systems, various aspects cannot be viewed in isolation. Often changes in one part of the overall system can propagate and cause changes in another part. Thus, you can only understand a problem if you look at its parts in relation to the whole. Based on this framework, IE looks at environmental issues with a *systems thinking* approach.

Take a city for instance. A city can be divided into commercial areas, residential areas, offices, services, infrastructures, and so forth. These are all sub-systems of the 'big city' system. Problems can emerge in one sub-system, but the solution has to be global. Let's say the price of housing is rising dramatically because there is too high a demand for housing. One solution would be to build new houses, but this will lead to more people living in the city, leading to the need for more infrastructure like roads, schools, more supermarkets, etc. This system is a simplified interpretation of reality whose behaviors can be 'predicted'.

In many cases, the systems IE deals with are complex systems. Complexity makes it difficult to understand the behavior of the system and may lead to rebound effects. Due to unforeseen behavioral change of users or consumers, a measure taken to improve environmental performance does not lead to any improvement or may even worsen the situation. For instance, in big cities, traffic can become problematic. Let's imagine the government wants to reduce air pollution and makes a policy stating that only cars with an even license plate number can drive on Tuesdays and Thursdays. Odd license plate numbers can drive on Wednesdays and Fridays. Finally, the other days, both cars are allowed on the roads. The first effect could be that people buy a second car, with a specific demand for license plate numbers, so they can drive every day. The rebound effect is that, the days when all cars are allowed to drive, some inhabitants now use both cars (whereas they only had one car to use before the policy). This policy example obviously did not lead to environmental improvement, but even made air pollution worse.

Moreover, *life cycle thinking* is also a very important principle in industrial ecology. It implies that all environmental impacts caused by a product, system, or project during its life cycle are taken into account. In this context life cycle includes

- Raw material extraction

- Material processing

- Manufacture

- Use

- Maintenance

- Disposal

The transport necessary between these stages is also taken into account as well as, if relevant, extra stages such as reuse, remanufacture, and recycle. Adopting a life cycle approach is essential to avoid shifting environmental impacts from one life cycle stage to another. This is commonly referred to as problem shifting. For instance, during the re-design of a product, one can choose to reduce its weight, thereby decreasing use of resources. However, it is possible that the lighter materials used in the new product will be more difficult to dispose of. The environmental impacts of the product gained during the extraction phase are shifted to the disposal phase. Overall environmental improvements are thus null.

A final and important principle of IE is its *integrated approach* or *multidisciplinarity*. IE takes into account three different disciplines: social sciences (including economics), technical sciences and environmental sciences. The challenge is to merge them into a single approach.

Tools

People	Planet	Profit	Modeling
<ul><li>Stakeholder analysis</li><li>Strength Weakness Opportunities Threats Analysis (SWOT Analysis)</li><li>Ecolabelling</li><li>ISO 14000</li><li>Environmental management system (EMS)</li><li>Integrated chain management (ICM)</li><li>Technology assessment</li></ul>	<ul><li>Environmental impact assessment (EIA)</li><li>Input-output analysis (IOA)</li><li>Life-cycle assessment (LCA)</li><li>Material flow analysis (MFA)</li><li>Substance flow analysis (SFA)</li><li>MET Matrix</li></ul>	<ul><li>Cost benefit analysis (CBA)</li><li>Full cost accounting (FCA)</li><li>Life cycle costing (LCC)</li></ul>	<ul><li>Stock and flow analysis</li><li>Agent based modeling</li></ul>

Criticisms

Industrial ecology implicitly assumes that technologies such as eco-efficiency, design for environment and material flow analysis can achieve an ecologically sustainable economy. This view has been recently challenged by Huesemann and Huesemann who claim that negative unintended consequences of technology are inherently unpredictable and unavoidable, that the techno-optimism reflected in industrial ecology is generally unjustified, and that modern technology, in the presence of continued economic growth, does not promote sustainability but rather hastens collapse. Therefore, the authors say, more than technological tinkering is needed to achieve long-term sustainability — human overpopulation must be reduced and a steady state economy is needed for sustainability.

Future Directions

The ecosystem metaphor popularized by Frosch and Gallopoulos has been a valuable creative tool for helping researchers look for novel solutions to difficult problems. Recently, it has been pointed out that this metaphor is based largely on a model of classical ecology, and that advancements in understanding ecology based on complexity science have been made by researchers such as C. S. Holling, James J. Kay, and others. For industrial ecology, this may mean a shift from a more mechanistic view of systems, to one where sustainability is viewed as an emergent property of a complex system. To explore this further, several researchers are working with agent based modeling techniques.

analysis is performed in the field of industrial ecology to use energy more efficiently. The term *exergy* was coined by Zoran Rant in 1956, but the concept was developed by J. Willard Gibbs. In recent decades, utilization of exergy has spread outside of physics and engineering to the fields of industrial ecology, ecological economics, systems ecology, and energetics.

Recently, there has been work advocating for large scale photovoltaic production facilities in an industrial ecology setting. These facilities not only reduce their environmental impact but also decrease the costs of photovoltaic productions to as little as $1 per Watt by economy of scale.

History of Industrial Ecology

The birth of industrial ecology is commonly attributed to an article devoted to industrial ecosystems, written by Frosch and Gallopoulos, which appeared in a 1989 special issue of Scientific American, but the field's fundamentals appeared much earlier. Industrial Ecology emerged from several ideas and concepts, some of which date back to the 19th century. Industrial ecology as a concept and as a field of scientific research has developed over time since.

Before the 1960S

The term "Industrial Ecology" has been used alongside "Industrial Symbiosis" at least since the 1940s. Economic geography was perhaps one of the first fields to use these terms. For example, in an article published in 1947, George T. Renner refers to "The General Principle of Industrial Location" as a "Law of Industrial Ecology". Briefly stated this is:

Any industry tends to locate at a point which provides optimum access to its ingredients or component elements. If all these component elements be juxtaposed, the location of the industry is predetermined. If, however, they occur widely separated, the industry is so located as to be most accessible to that element which would be the most expensive or difficult to transport and which, therefore, becomes the locative factor for the industry in question.

In the same article the author defines and describes industrial symbiosis:

Often the location of an industry cannot be fully understood solely in terms of its locative ingredient elements. There are relationships between industries, sometimes simple, but often quite complex, which enter into and complicate the analysis. Chief among these is the phenomenon of industrial symbiosis. By this is meant the consorting together of two or more of dissimilar industries. Industrial Symbiosis, when scrutinized, is seen to be of two kinds, disjunctive and conjunctive.

It appears that the concept of Industrial Symbiosis was not new for the field of economic geography, since the same categorization is used by Walter G. Lezius in his 1937 article "Geography of Glass Manufacture at Toledo, Ohio", also published in the Journal of Economic Geography.

Used in a different context, the term "Industrial Ecology" is also found in a 1958 paper concerned with the relationship between the ecological impact from increasing urbanization and value orientations of related peoples. The case study is in Lebanon:

The central ecological variable in the present research is ecological mobility, or the movement of men in space. It is patent that modern Industrial Ecology requires more such adaptive mobility than does traditional folk-village organization.

1960s

In 1963, we find the term Industrial Ecology (defined as the "complex ecology of the modern industrial world") being used to describe the social nature and complexity of (and within) industrial systems:

...industrial organisations are social rather than mechanical systems. A firm is not only a working organisation with a working purpose. It is rather a community with its own 'politics', in so far as it is involved in problems concerned with the proper distribution of power between individuals and groups of individuals and with questions of individual and group prestige, influence, status and standing... [and he concludes that] the understanding which the student of management is expected to gain is no less than the attainment of insight into an Industrial Ecology of great complexity.

In 1967, the President of the American association for the advancement of science writes in "The experimental city" that "There are examples of industrial symbiosis where one industry feeds off, or at least neutralizes, the wastes of another..." The same author in 1970 talks about "The Next Industrial Revolution" The concept of material and energy sharing and reuse is central to his proposal for a new industrial revolution and he cites agro-industrial symbiosis as a practical way for achieving this:

The object of the next industrial revolution is to ensure that there will be no such thing as waste, on the basis that waste is simply some substance that we do not yet have the wit to use... The next industrial revolution is this generating of a huge new [industry that]... will not produce products, it will rather reprocess the things we call wastes so they may be reproduced in the factories into the things we need... Having the city near the rural area will enable waste heat to be used to speed up the biological processes of treating the organic wastes before they go back into the land. This might end in an elegant arrangement-the power plants located close enough to the center of use, to the people who need the power, but also, within the economics, close enough to the agriculture lands so that the waste heat may be used there. This is an example of agro-industrial symbiosis, if you like to call it that.

In these early articles, "Industrial Ecology" is used in its literal sense - as a system of interacting industrial entities. The relation to natural ecosystems (through either metaphor or analogy) is not explicit. Industrial Symbiosis on the other hand, is already clearly defined as a type of industrial organization, and the term symbiosis is borrowed from the ecological sciences to describe an analogous phenomenon in industrial systems.

1970s

Industrial Ecology has been a research subject of the Japan Industrial Policy Research Institute since 1971. Their definition of Industrial Ecology is "research for the prospect of dynamic harmonization between human activities and nature by a systems approach based upon ecology (JIPRI, 1983)". This programme has resulted to a number of reports that are available only in Japanese.

One of the earliest definitions of Industrial Ecology was proposed by Harry Zvi Evan at a seminar of the Economic Commission of Europe in Warsaw (Poland) in 1973 (an article was subsequently published by Evan in the Journal for International Labour Review in 1974 vol. 110 (3), pp. 219–233). Evan defined Industrial Ecology as a systematic analysis of industrial operations including factors like: Technology, environment, natural resources, bio-medical aspects, institutional and legal matters as well as the socio-economic aspects.

In 1974 the term of Industrial Ecology is perhaps for the first time associated with a cyclical production mode (rather than a linear one, resulting to waste). In this article, the necessity for a transition to an "open-world Industrial Ecology", is used as argument for the need to establish lunar industries:

Low living standards provide one strong motive for most developing countries to increase their productivity and grow economically. Population increase (while it lasts) is a still more powerful driver for increased world consumption. Thus the pressure on resources will continue to grow. Instead of deploring it, we better grow with it. Only through transition to an open-world Industrial Ecology - which includes both benign industrial revolution on Earth and extraterrestrial industrialization - can the present apparent limits to growth be overcome.

Many elements of modern Industrial Ecology were commonplace in the industrial sectors of the former Soviet Union. For example, "kombinirovanaia produksia" (combined production) was present from the earliest years of the Soviet Union and was instrumental in shaping the patterns of Soviet industrialization. "Bezotkhodnoyi tekhnologii" (waste-free technology) was introduced in the final decades of the USSR as a way to increase industrial production while limiting environmental impact. Fiodor Davitaya, a Soviet scientist from the Republic of Georgia, described in 1977 the analogy relating industrial systems to natural systems as a model for a desirable transition to cleaner production:

Nature operates without any waste products. What is rejected by some organisms provides food for others. The organisation of industry on this principle—with the waste products of some branches of industry providing raw material for others—means in effect using natural processes as a model, for in them the resolution of all arising contradictions is the motive force of progress.

1980s

By the 80s Industrial Ecology was already "promoted" to a research subject, which several institutes around the globe embraced. In a 1986 article published in the Journal of Ecological Modeling, there is a full description of Industrial Ecology and the analogy to natural ecosystems is clearly stated:

The structure and inner-working of an industrial society resemble those of a natural ecosystem. The concepts in ecology such as habitat, succession, trophic level, limiting factors and community metabolism can also apply to the study of the ecology of an industrial society. For instance, an industry in a society may grow or decline as a consequence of dynamic changes in exogenous limiting resources and in the hierarchical and/or metabolic structure of that society. When studying the ecology of an industrial society (henceforth termed 'Industrial Ecology'), these concepts and methodologies employed in ecosystems analyses are useful.

In fact, in the above article there is an attempt to model an "industrial ecological system". The model is composed of seven major sections: industry, population, labor force, living state, environment and pollution, general health, and occupational health. Notice the rough similarity with Evan's factors as stated in the above section.

During the 80s the emergence of another related term, "industrial metabolism", is observed. The term is used as a metaphor for the organization and functioning of industrial activity. In an article defending the "biological modulation of terrestrial carbon cycle", the author includes an extraordinary parenthetical note:

Parenthetically, it should be noted that it is an intrinsic property of life to proliferate exponentially until the encounter of limits set by (1) the availability of biologically utilizable reducing power, or (2) the exhaustion of some critical nutrient, or (3) an autotoxic effect imposed by life on its own environment. These limits are universal, applying to microbial ecosystems as well as to the population dynamics of a seemingly unrestricted biological superdominant such as Homo Sapiens (here, the ultimate limit is likely to be placed by an autotoxic effect exerted by the "extrasomatic" (industrial) metabolism of the human race).

1989 – Decisive Articles

In 1989 two articles were released that played a decisive role in the history of industrial ecology. The first one was titled "Industrial Metabolism" by Robert Ayres. Ayres essentially lays the foundations of Industrial Ecology, although the term is not to be found in this article. In the appendix of the article he includes "a theoretical exploration of the biosphere and the industrial economy

as material-transformation systems and lessons that might be learned from their comparison".
He proposes that:

We may think of both the biosphere and the industrial economy as systems for the transformation
of materials. The biosphere as it now exists is nearly a perfect system for recycling materials. This
was not the case when life on earth began. The industrial system of today resembles the earliest
stage of biological evolution, when the most primitive living organisms obtained their energy from
a stock of organic molecules accumulated during prebiotic times. It is increasingly urgent for us to
learn from the biosphere and modify our industrial metabolism, the energy - and value - yielding
process essential to economic development... we should not only postulate, but indeed endorse, a
long-run imperative favoring an industrial metabolism that results in reduced extraction of virgin
materials, reduced loss of waste materials, and increased recycling of useful ones.

The term "Industrial Ecology" gains mainstream attention later the same year (1989) through a
"Scientific American" article named "Strategies for Manufacturing". In this article, R.Frosch and
N.Gallopoulos wonder "why would not our industrial system behave like an ecosystem, where the
wastes of a species may be resource to another species? Why would not the outputs of an industry be the inputs of another, thus reducing use of raw materials, pollution, and saving on waste
treatment?"

This vision gave birth to the concept of the Eco-industrial Park, the industrial complex that is
governed by Industrial Ecology principles. A notable example resides in a Danish industrial park
in the city of Kalundborg. There, several linkages of byproducts and waste heat can be found between numerous entities such as a large power plant, an oil refinery, a pharmaceutical plant, a
plasterboard factory, an enzyme manufacturer, a waste company and the city itself.

and Gallopoulos' thinking was in certain ways simply an extension of earlier ideas, such as the ef-
ficiency and waste-reduction thinking annunciated by Buckminster Fuller and his students (e.g.,
J. Baldwin), and parallel ideas about energy cogeneration, such as those of Amory Lovins and the
Rocky Mountain Institute.

1990s

In 1991, C. Kumar Patel organized a seminal colloquium on Industrial Ecology, held on May 20
and 21, 1991, at the National Academy of Sciences in Washington D.C. The papers were later published in the Proceedings of the National Academy of Sciences USA, and they form an excellent
reference on Industrial Ecology. Papers include

- "Industrial Ecology: Concepts and Approaches"
- "Industrial Ecology: A Philosophical Introduction"
- "The Ecology of Markets,"
- "Industrial Ecology: Reflections on a Colloquium"

21St Century

The scientific field Industrial Ecology has grown fast in recent years. The Journal of Industrial

Ecology (since 1997), the International Society for Industrial Ecology (since 2001), and the journal Progress in Industrial Ecology (since 2004) give Industrial Ecology a strong and dynamic position in the international scientific community. Industrial Ecology principles are also emerging in various policy realms such as the concept of the Circular Economy that is being promoted in China. Although the definition of the Circular Economy has yet to be formalized, generally the focus is on strategies such as creating a circular flow of materials, and cascading energy flows. An example of this would be using waste heat from one process to run another process that requires a lower temperature. This maximizes the efficiency of exergy use. The hope is that strategy such as this will create a more efficient economy with fewer pollutants and other unwanted by products.

References

- Huesemann, Michael H., and Joyce A. Huesemann (2011). Technofix: Why Technology Won't Save Us or the Environment, New Society Publishers, Gabriola Island, British Columbia, Canada, ISBN 0865717044, 464 pp.

9

Various Industrial Systems

Industries are an amalgamation of systems, in order to understand industries as a whole, it is essential that the readers develop a comprehensive understanding about its different systems. This chapter will focus on the various industrial systems like pilot plant, demonstration plant and industrial robot. The chapter will be an invaluable source for readers as it will broaden their knowledge of this area of study.

Pilot Plant

A pilot plant is a small industrial system, which is operated to generate information about the behavior of the system for use in design of larger facilities. Pilot plant is a relative term in the sense that plants are typically smaller than full-scale production plants, but are built in a range of sizes. Some pilot plants are built in laboratories using stock lab equipment, while others require substantial engineering efforts, cost millions of dollars, and are custom-assembled and fabricated from process equipment, instrumentation and piping. They can also be used to train personnel for a full-scale plant. Pilot plants tend to be smaller compared to demonstration plants.

A pilot plant of a bioreactor

Risk Management

Pilot plants are used to reduce the risk associated with construction of large process plants. They do so in several ways:

- Computer simulations and semi-empirical methods are used to determine the limitations of the pilot scale system. These mathematical models are then tested in a physical pilot-scale plant. Various modeling methods are used for scale-up. These methods include:

  o Chemical similitude studies

  o Mathematical modeling

  o Aspen Plus/Aspen HYSYS modeling

  o Finite Elemental Analysis (FEA)

  o Computational Fluid Dynamics (CFD)

    - These theoretical modeling methods return the following:

      - Finalized mass and energy balances

      - Optimized system design and capacity

      - Equipment requirements

      - System limitations

      - The basis for determining the cost to build the pilot module

- They are substantially less expensive to build than full-scale plants. The business does not put as much capital at risk on a project that may be inefficient or unfeasible. Further, design changes can be made more cheaply at the pilot scale and kinks in the process can be worked out before the large plant is constructed.

- They provide valuable data for design of the full-scale plant. Scientific data about reactions, material properties, corrosiveness, for instance, may be available, but it is difficult to predict the behavior of a process of any complexity. Engineering data from other process may be available, but this data can not always be clearly applied to the process of interest. Designers use data from the pilot plant to refine their design of the production scale facility.

If a system is well defined and the engineering parameters are known, pilot plants are not used. For instance, a business that wants to expand production capacity by building a new plant that does the same thing as an existing plant may choose to not use a pilot plant.

Additionally, advances in process simulation on computers have increased the confidence of process designers and reduced the need for pilot plants. However, they are still used as even state-of-the-art simulation cannot accurately predict the behavior of complex systems.

Design Change

As a system increases in size, system properties that depend on quantity of matter (with extensive properties) might change. The chemical and physical properties of a system affect each other and create varying results. The surface area to liquid ratio is a good example of such a property. On a small chemical scale, in a flask, say, there is a relatively large surface area to liquid ratio. However, if the reaction in question is scaled up to fit in a 500-gallon tank, the surface area to liquid ratio becomes much smaller. As a result of this difference in surface area to liquid ratio, the exact nature of the thermodynamics and the reaction kinetics of the process changes in a non-linear fashion. This is why a reaction in a beaker can behave vastly differently from the same reaction in a large-scale production process.

Other Factors

Other factors that change during the transformation to a production scale include:

- Reaction kinetics

- Chemical equilibrium

- Material properties

- Fluid dynamics

- Thermodynamics

- Equipment selection

- Agitation

After data is collected from operation of a pilot plant, a larger production-scale facility may be built. Alternatively, a demonstration plant, which is bigger than a pilot plant, but smaller than the full-scale production plant, may be built to demonstrate the commercial feasibility of the process. Businesses sometimes continue to operate the pilot plant in order to test ideas for new products, new feedstocks, or different operating conditions. Alternatively, they may be operated as production facilities, augmenting production from the main plant.

Recent trends try to keep the size of the plant a small as possible to save costs. This approach is called miniplant technology. The flow chemistry takes up this trend and uses flow miniplant technology for small-scale manufacturing.

Bench Scale Vs Pilot Vs Demonstration

The differences between bench scale, pilot scale and demonstration scale are strongly influenced by industry and application. Some industries use pilot plant and demonstration plant interchangeably. Some pilot plants are built as portable modules that can easily transported as a contained unit.

For batch processes, in the pharmaceutical industry for example, bench scale is typically conducted on samples 1–20 kg or less, whereas pilot scale testing is performed with samples of 20–100 kg. Demonstration scale is essentially operating the equipment at full commercial feed rates over extended time periods to prove operational stability.

For continuous processes, in the petroleum industry for example, bench scale systems are typically microreactor or CSTR systems with less than 1000 cc of catalyst, studying reactions and/or separations on a once-through basis. Pilot plants will typically have reactors with catalyst volume between 1 and 100 litres filled with Jesse, and will often incorporate product separation and gas/liquid recycle with the goal of closing the mass balance. Demonstration plants, also referred to as semi-works plants, will study the viability of the process on a pre-commercial scale, with typical catalyst volumes in the 100 - 1000 litre range. The design of a demonstration scale plant for a continuous process will closely resemble that of the anticipated future commercial plant, albeit at a much lower throughput, and its goal is to study catalyst performance and operating lifetime over an extended period, while generating significant quantities of product for market testing.

In the development of new processes, the design and operation of the pilot and demonstration plant will often run in parallel with the design of the future commercial plant, and the results from pilot testing programs are key to optimizing the commercial plant flowsheet. It is common in cases where process technology has been successfully implemented that the savings at the commercial scale resulting from pilot testing will significantly outweigh the cost of the pilot plant itself.

Steps to Creating a Custom Pilot Plant

Custom pilot plants are commonly designed either for research or commercial purposes. They can range in size from a small system with no automation and low flow, to a highly automated system producing relatively large amounts of products in a day. No matter the size, the steps to designing and fabricating a working pilot plant are the same. They are:

1. Pre-engineering - completing a process flow diagram (PFD), basic piping and instrumentation diagrams (P&ID's) and initial equipment layouts.

2. Engineering modeling and optimization - 2D and 3D models are created, using a simulation software to model the process parameters and scale the chemical processes. These modeling softwares help determine system limitations, non-linear chemical and physical changes, and potential equipment sizing. Mass and energy balances, finalized P&ID's and general arrangement drawings are produced.

3. Automation strategies for the system are developed (if needed). Controls system programming begins and will continue through fabrication and assembly

4. Fabrication and assembly - after an optimized design has been determined, the custom pilot is fabricated and assembled. Pilot plants can either be assembled on-site or off-site as modular skids that will be constructed and tested in a controlled environment.

5. Testing - testing of completed systems, including system controls, is conducted to ensure proper system function.

6. Installation and startup - if constructed offsite, pilot skids are installed onsite. After all equipment is in place, full system startup is completed by integrating the system with existing plant utilities and controls. Full operation is tested and affirmed.

7. Training - operator training is complete and full system documentation is handed over.

Bibliography

- M. Levin (Editor), Pharmaceutical Process Scale-Up (Drugs and the Pharmaceutical), Informa Healthcare, 3rd edition, ISBN 978-1616310011 (2011)

- M. Lackner (Editor), Scale-up in Combustion, ProcessEng Engineering GmbH, Wien, ISBN 978-3-902655-04-2 (2009).

- M. Zlokarnik, Scale-up in Chemical Engineering, Wiley-VCH Verlag GmbH & Co. KGaA, 2nd edition, ISBN 978-3527314218 (2006).

Richard Palluzi, Pilot Plants: Design, Construction and Operation, McGraw-Hill, February, 1992.
Richard Palluzi, Pilot Plants, Chemical Engineering, March, 1990.

Demonstration Plant

Demonstration Plant

A demonstration plant is an industrial system used to validate an industrial process for commercialization. It is larger than a pilot plant, and is the final stage in research, development and demonstration of a new process. Demonstration plants are built in a range of sizes, and the term 'demonstration plant' can sometimes be used interchangeably with 'pilot plant.' However, demonstration plants are generally larger than pilot plants are often constructed following a successful trial in a pilot scale size. Demonstration plants are used to prove a process works at industrial scale, and is financially viable in its intended industry.

Goals

The goals of a demonstration plant are generally as follows:

- Prove a new technology using commercially available, pre-tested equipment.

- Show a reasonable return on investment ROI for the capital that will be invested in a full-scale system, including the operational costs of running such a system.

- In some case, to start bringing product to market in significant enough amounts that production, distribution and target market viability can be established, including finalization of market testing.

- Establish a viable product method that will endure the test of a true manufacturing operation.

Design Factors

Many of the same design techniques that are used for pilot plants are also used when developing demonstration plants. 3D Modeling, chemical similitude studies, mass and energy balances, risk factors, computational fluid dynamics (CFD), and mathematical modeling are common techniques used to design demonstration modules before actual fabrication occurs.

The emphasis in a demonstration plant is on using industrial equipment, rather than smaller scale equipment, to prove process viability. A significant amount of product must be produced in equipment that will hold up over a long production life-time and not be prohibitively expensive. A demonstration plant must show that enough end-product can be created to off-set the costs of the commercial system over a period of time.

Industrial Robot

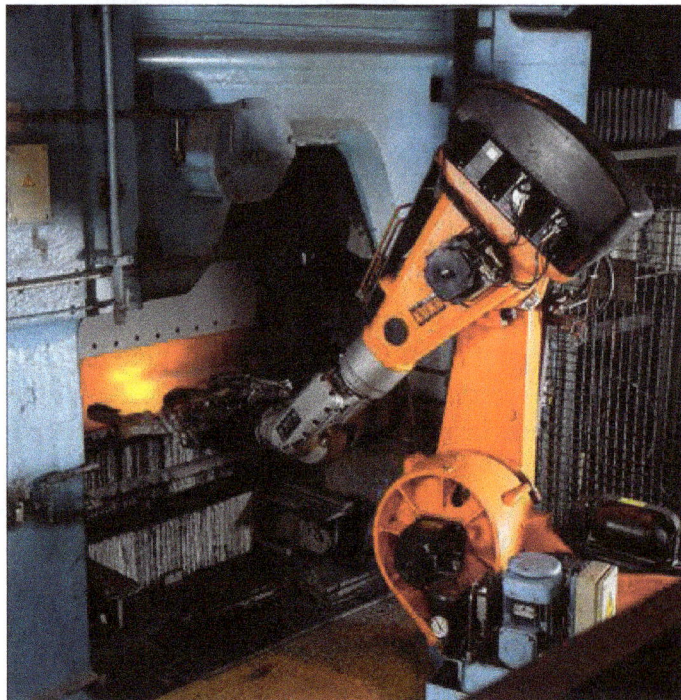

Articulated industrial robot operating in a foundry.

An industrial robot is a robot system used for manufacturing. Industrial robots are automated, programmable and capable of movement on two or more axes.

Typical applications of robots include welding, painting, assembly, pick and place for printed circuit boards, packaging and labeling, palletizing, product inspection, and testing; all accomplished with high endurance, speed, and precision. They can help in material handling and provide interfaces.

Types and Features

A set of six-axis robots used for welding.

Factory Automation with industrial robots for palletizing food products like bread and toast at a bakery in Germany

The most commonly used robot configurations are articulated robots, SCARA robots, delta robots and cartesian coordinate robots, (gantry robots or x-y-z robots). In the context of general robotics, most types of robots would fall into the category of robotic arms (inherent in the use of the word *manipulator* in ISO standard 1738). Robots exhibit varying degrees of autonomy:

- Some robots are programmed to faithfully carry out specific actions over and over again (repetitive actions) without variation and with a high degree of accuracy. These actions are determined by programmed routines that specify the direction, acceleration, velocity, deceleration, and distance of a series of coordinated motions.

- Other robots are much more flexible as to the orientation of the object on which they are operating or even the task that has to be performed on the object itself, which the robot may even need to identify. For example, for more precise guidance, robots often contain machine vision sub-systems acting as their visual sensors, linked to powerful computers or controllers. Artificial intelligence, or what passes for it, is becoming an increasingly important factor in the modern industrial robot.

History of Industrial Robotics

The earliest known industrial robot, conforming to the ISO definition was completed by "Bill" Griffith P. Taylor in 1937 and published in Meccano Magazine, March 1938. The crane-like device was built almost entirely using Meccano parts, and powered by a single electric motor. Five axes of movement were possible, including *grab* and *grab rotation*. Automation was achieved using punched paper tape to energise solenoids, which would facilitate the movement of the crane's control levers. The robot could stack wooden blocks in pre-programmed patterns. The number of motor revolutions required for each desired movement was first plotted on graph paper. This information was then transferred to the paper tape, which was also driven by the robot's single motor. Chris Shute built a complete replica of the robot in 1997.

George Devol, c. 1982

George Devol applied for the first robotics patents in 1954 (granted in 1961). The first company to produce a robot was Unimation, founded by Devol and Joseph F. Engelberger in 1956, and the first robot, Unimate, was based on Devol's original patents. Unimation robots were also called *programmable transfer machines* since their main use at first was to transfer objects from one point to another, less than a dozen feet or so apart. They used hydraulic actuators and were programmed in *joint coordinates*, i.e. the angles of the various joints were stored during a teaching phase and replayed in operation. They were accurate to within 1/10,000 of an inch. Unimation later licensed their technology to Kawasaki Heavy Industries and GKN, manufacturing Unimates in Japan and England respectively. For some time Unimation's only competitor was Cincinnati Milacron Inc. of Ohio. This changed radically in the late 1970s when several big Japanese conglomerates began producing similar industrial robots.

In 1969 Victor Scheinman at Stanford University invented the Stanford arm, an all-electric, 6-axis articulated robot designed to permit an arm solution. This allowed it accurately to follow arbitrary paths in space and widened the potential use of the robot to more sophisticated applications such as assembly and welding. Scheinman then designed a second arm for the MIT AI Lab, called the "MIT arm." Scheinman, after receiving a fellowship from Unimation to develop his designs, sold those designs to Unimation who further developed them with support from General Motors and later marketed it as the Programmable Universal Machine for Assembly (PUMA).

Industrial robotics took off quite quickly in Europe, with both ABB Robotics and KUKA Robotics bringing robots to the market in 1973. ABB Robotics (formerly ASEA) introduced IRB 6, among the world's first *commercially available* all electric micro-processor controlled robot. The first two IRB 6 robots were sold to Magnusson in Sweden for grinding and polishing pipe bends and were installed in production in January 1974. Also in 1973 KUKA Robotics built its first robot, known as FAMULUS, also one of the first articulated robots to have six electromechanically driven axes.

Interest in robotics increased in the late 1970s and many US companies entered the field, including large firms like General Electric, and General Motors (which formed joint venture FANUC Robotics with FANUC LTD of Japan). U.S. startup companies included Automatix and Adept Technology, Inc. At the height of the robot boom in 1984, Unimation was acquired by Westinghouse Electric Corporation for 107 million U.S. dollars. Westinghouse sold Unimation to Stäubli Faverges SCA of France in 1988, which is still making articulated robots for general industrial and cleanroom applications and even bought the robotic division of Bosch in late 2004.

Only a few non-Japanese companies ultimately managed to survive in this market, the major ones being: Adept Technology, Stäubli-Unimation, the Swedish-Swiss company ABB Asea Brown Boveri, the German company KUKA Robotics and the Italian company Comau.

Technical Description

Defining Parameters

- *Number of axes* – two axes are required to reach any point in a plane; three axes are required to reach any point in space. To fully control the orientation of the end of the arm (i.e. the *wrist*) three more axes (yaw, pitch, and roll) are required. Some designs (e.g. the SCARA robot) trade limitations in motion possibilities for cost, speed, and accuracy.

- *Degrees of freedom* – this is usually the same as the number of axes.

- *Working envelope* – the region of space a robot can reach.

- *Kinematics* – the actual arrangement of rigid members and joints in the robot, which determines the robot's possible motions. Classes of robot kinematics include articulated, cartesian, parallel and SCARA.

- *Carrying capacity or payload* – how much weight a robot can lift.

- *Speed* – how fast the robot can position the end of its arm. This may be defined in terms of the angular or linear speed of each axis or as a compound speed i.e. the speed of the end of the arm when all axes are moving.

- *Acceleration* – how quickly an axis can accelerate. Since this is a limiting factor a robot may not be able to reach its specified maximum speed for movements over a short distance or a complex path requiring frequent changes of direction.

- *Accuracy* – how closely a robot can reach a commanded position. When the absolute position of the robot is measured and compared to the commanded position the error is a measure of accuracy. Accuracy can be improved with external sensing for example a vision system or Infra-Red. Accuracy can vary with speed and position within the working envelope and with payload.

- *Repeatability* – how well the robot will return to a programmed position. This is not the same as accuracy. It may be that when told to go to a certain X-Y-Z position that it gets only to within 1 mm of that position. This would be its accuracy which may be improved by calibration. But if that position is taught into controller memory and each time it is sent there it returns to within 0.1mm of the taught position then the repeatability will be within 0.1mm.

Accuracy and repeatability are different measures. Repeatability is usually the most important criterion for a robot and is similar to the concept of 'precision' in measurement. ISO 9283 sets out a method whereby both accuracy and repeatability can be measured. Typically a robot is sent to a taught position a number of times and the error is measured at each return to the position after visiting 4 other positions. Repeatability is then quantified using the standard deviation of those samples in all three dimensions. A typical robot can, of course make a positional error exceeding that and that could be a problem for the process. Moreover, the repeatability is different in different parts of the working envelope and also changes with speed and payload. ISO 9283 specifies that accuracy and repeatability should be measured at maximum speed and at maximum payload. But this results in pessimistic values whereas the robot could be much more accurate and repeatable at light loads and speeds. Repeatability in an industrial process is also subject to the accuracy of the end effector, for example a gripper, and even to the design of the 'fingers' that match the gripper to the object being grasped. For example, if a robot picks a screw by its head, the screw could be at a random angle. A subsequent attempt to insert the screw into a hole could easily fail. These and similar scenarios can be improved with 'lead-ins' e.g. by making the entrance to the hole tapered.

- *Motion control* – for some applications, such as simple pick-and-place assembly, the robot need merely return repeatably to a limited number of pre-taught positions. For more sophisticated applications, such as welding and finishing (spray painting), motion must be continuously controlled to follow a path in space, with controlled orientation and velocity.

- *Power source* – some robots use electric motors, others use hydraulic actuators. The former are faster, the latter are stronger and advantageous in applications such as spray painting, where a spark could set off an explosion; however, low internal air-pressurisation of the arm can prevent ingress of flammable vapours as well as other contaminants.

- *Drive* – some robots connect electric motors to the joints via gears; others connect the motor to the joint directly (*direct drive*). Using gears results in measurable 'backlash' which is free movement in an axis. Smaller robot arms frequently employ high speed, low torque DC motors, which generally require high gearing ratios; this has the disadvantage of backlash. In such cases the harmonic drive is often used.

- *Compliance* - this is a measure of the amount in angle or distance that a robot axis will move when a force is applied to it. Because of compliance when a robot goes to a position carrying its maximum payload it will be at a position slightly lower than when it is carrying no payload. Compliance can also be responsible for overshoot when carrying high payloads in which case acceleration would need to be reduced.

Robot Programming and Interfaces

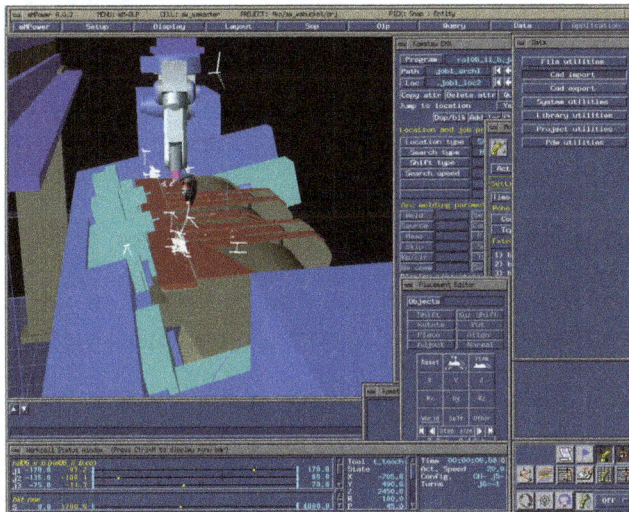

Offline programming by ROBCAD

A typical well-used teach pendant with optional mouse

The setup or programming of motions and sequences for an industrial robot is typically taught by linking the robot controller to a laptop, desktop computer or (internal or Internet) network.

A robot and a collection of machines or peripherals is referred to as a workcell, or cell. A typical cell might contain a parts feeder, a molding machine and a robot. The various machines are 'integrated' and controlled by a single computer or PLC. How the robot interacts with other machines in the cell must be programmed, both with regard to their positions in the cell and synchronizing with them.

Software: The computer is installed with corresponding interface software. The use of a computer greatly simplifies the programming process. Specialized robot software is run either in the robot controller or in the computer or both depending on the system design.

There are two basic entities that need to be taught (or programmed): positional data and procedure. For example, in a task to move a screw from a feeder to a hole the positions of the feeder and the hole must first be taught or programmed. Secondly the procedure to get the screw from the feeder to the hole must be programmed along with any I/O involved, for example a signal to indicate when the screw is in the feeder ready to be picked up. The purpose of the robot software is to facilitate both these programming tasks.

Teaching the robot positions may be achieved a number of ways:

Positional commands The robot can be directed to the required position using a GUI or text based commands in which the required X-Y-Z position may be specified and edited.

Teach pendant: Robot positions can be taught via a teach pendant. This is a handheld control and programming unit. The common features of such units are the ability to manually send the robot to a desired position, or "inch" or "jog" to adjust a position. They also have a means to change the speed since a low speed is usually required for careful positioning, or while test-running through a new or modified routine. A large emergency stop button is usually included as well. Typically once the robot has been programmed there is no more use for the teach pendant.

Lead-by-the-nose is a technique offered by many robot manufacturers. In this method, one user holds the robot's manipulator, while another person enters a command which de-energizes the robot causing it to go limp. The user then moves the robot by hand to the required positions and/or along a required path while the software logs these positions into memory. The program can later run the robot to these positions or along the taught path. This technique is popular for tasks such as paint spraying.

Offline programming is where the entire cell, the robot and all the machines or instruments in the workspace are mapped graphically. The robot can then be moved on screen and the process simulated. A robotics simulator is used to create embedded applications for a robot, without depending on the physical operation of the robot arm and end effector. The advantages of robotics simulation is that it saves time in the design of robotics applications. It can also increase the level of safety associated with robotic equipment since various "what if" scenarios can be tried and tested before the system is activated. Robot simulation software provides a platform to teach, test, run, and debug programs that have been written in a variety of programming languages.

Robot simulation tools allow for robotics programs to be conveniently written and debugged off-line with the final version of the program tested on an actual robot. The ability to preview the behavior of a robotic system in a virtual world allows for a variety of mechanisms, devices, configurations and controllers to be tried and tested before being applied to a "real world" system. Robotics simulators have the ability to provide real-time computing of the simulated motion of an industrial robot using both geometric modeling and kinematics modeling.

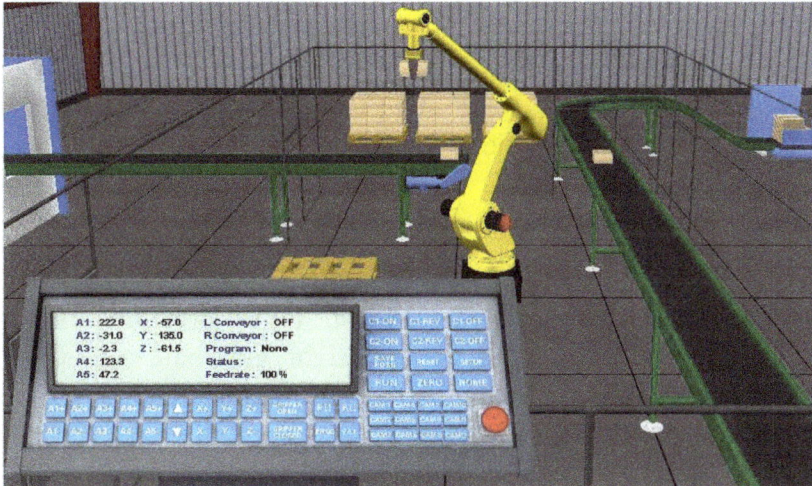

RoboLogix Robotics Simulator.

Others In addition, machine operators often use user interface devices, typically touchscreen units, which serve as the operator control panel. The operator can switch from program to program, make adjustments within a program and also operate a host of peripheral devices that may be integrated within the same robotic system. These include end effectors, feeders that supply components to the robot, conveyor belts, emergency stop controls, machine vision systems, safety interlock systems, bar code printers and an almost infinite array of other industrial devices which are accessed and controlled via the operator control panel.

The teach pendant or PC is usually disconnected after programming and the robot then runs on the program that has been installed in its controller. However a computer is often used to 'supervise' the robot and any peripherals, or to provide additional storage for access to numerous complex paths and routines.

End-of-Arm Tooling

The most essential robot peripheral is the end effector, or end-of-arm-tooling (EOT). Common examples of end effectors include welding devices (such as MIG-welding guns, spot-welders, etc.), spray guns and also grinding and deburring devices (such as pneumatic disk or belt grinders, burrs, etc.), and grippers (devices that can grasp an object, usually electromechanical or pneumatic). Another common means of picking up an object is by vacuum. End effectors are frequently highly complex, made to match the handled product and often capable of picking up an array of products at one time. They may utilize various sensors to aid the robot system in locating, handling, and positioning products.

Controlling Movement

For a given robot the only parameters necessary to completely locate the end effector (gripper, welding torch, etc.) of the robot are the angles of each of the joints or displacements of the linear axes (or combinations of the two for robot formats such as SCARA). However, there are many different ways to define the points. The most common and most convenient way of defining a point is to specify a Cartesian coordinate for it, i.e. the position of the 'end effector' in mm in the X, Y and Z directions relative to the robot's origin. In addition, depending on the types of joints a particular robot may have, the orientation of the end effector in yaw, pitch, and roll and the location of the tool point relative to the robot's faceplate must also be specified. For a jointed arm these coordinates must be converted to joint angles by the robot controller and such conversions are known as Cartesian Transformations which may need to be performed iteratively or recursively for a multiple axis robot. The mathematics of the relationship between joint angles and actual spatial coordinates is called kinematics.

Positioning by Cartesian coordinates may be done by entering the coordinates into the system or by using a teach pendant which moves the robot in X-Y-Z directions. It is much easier for a human operator to visualize motions up/down, left/right, etc. than to move each joint one at a time. When the desired position is reached it is then defined in some way particular to the robot software in use, e.g. P1 - P5 below.

Typical Programming

Most articulated robots perform by storing a series of positions in memory, and moving to them at various times in their programming sequence. For example, a robot which is moving items from one place to another might have a simple 'pick and place' program similar to the following:

Define points P1–P5:

1. Safely above workpiece (defined as P1)

2. 10 cm Above bin A (defined as P2)

3. At position to take part from bin A (defined as P3)

4. 10 cm Above bin B (defined as P4)

5. At position to take part from bin B. (defined as P5)

Define program:

1. Move to P1

2. Move to P2

3. Move to P3

4. Close gripper

5. Move to P2

6. Move to P4

7. Move to P5

8. Open gripper

9. Move to P4

10. Move to P1 and finish

Singularities

The American National Standard for Industrial Robots and Robot Systems — Safety Requirements (ANSI/RIA R15.06-1999) defines a singularity as "a condition caused by the collinear
alignment of two or more robot axes resulting in unpredictable robot motion and velocities." It
is most common in robot arms that utilize a "triple-roll wrist". This is a wrist about which the
three axes of the wrist, controlling yaw, pitch, and roll, all pass through a common point. An
example of a wrist singularity is when the path through which the robot is traveling causes the
first and third axes of the robot's wrist (i.e. robot's axes 4 and 6) to line up. The second wrist axis
then attempts to spin 180° in zero time to maintain the orientation of the end effector. Another
common term for this singularity is a "wrist flip". The result of a singularity can be quite dramatic
and can have adverse effects on the robot arm, the end effector, and the process. Some industrial
robot manufacturers have attempted to side-step the situation by slightly altering the robot's
path to prevent this condition. Another method is to slow the robot's travel speed, thus reducing
the speed required for the wrist to make the transition. The ANSI/RIA has mandated that robot
manufacturers shall make the user aware of singularities if they occur while the system is being
manually manipulated.

A second type of singularity in wrist-partitioned vertically articulated six-axis robots occurs when
the wrist center lies on a cylinder that is centered about axis 1 and with radius equal to the distance between axes 1 and 4. This is called a shoulder singularity. Some robot manufacturers also
mention alignment singularities, where axes 1 and 6 become coincident. This is simply a sub-case
of shoulder singularities. When the robot passes close to a shoulder singularity, joint 1 spins very
fast.

The third and last type of singularity in wrist-partitioned vertically articulated six-axis robots
occurs when the wrist's center lies in the same plane as axes 2 and 3.

Singularities are closely related to the phenomena of Gimbal Lock, which has a similar root cause
of axes becoming lined up.

A video illustrating these three types of singular configurations is available here.

Market Structure

According to the International Federation of Robotics (IFR) study *World Robotics 2014*, there
were between 1,332,000 and 1,600,000 operational industrial robots by the end of 2013. This
number is estimated to reach 1,946,000 by the end of 2017.

For the year 2011 the IFR estimates the worldwide sales of industrial robots with US$8.5 billion.

Including the cost of software, peripherals and systems engineering, the annual turnover for robot systems is estimated to be US$25.5 billion in 2011.

The Japanese government estimates the industry could surge from about $5.2 billion in 2006 to $26 billion in 2010 and nearly $70 billion by 2025. In 2005, there were over 370,000 operational industrial robots in Japan. A 2007 national technology roadmap by the Trade Ministry calls for 1 million industrial robots to be installed throughout the country by 2025.

Estimated Worldwide Annual Supply of Industrial Robots (in Units):

Year	supply
1998	69,000
1999	79,000
2000	99,000
2001	78,000
2002	69,000
2003	81,000
2004	97,000
2005	120,000
2006	112,000
2007	114,000
2008	113,000
2009	60,000
2010	118,000
2012	159,346
2013	178,132
2014 (forecast)	205,000

References

- Schott, John. "PILOT PLANT DESIGN AND CONSTRUCTION". EPIC Modular Process. EPIC Systems, Inc. Retrieved 31 May 2016.

- Gertenbach, Dennis; Cooper, Brian. "SCALEUP ISSUES FROM BENCH TO PILOT" (PDF). Hazen Research. AIChE.Retrieved 28 March 2016.

- Alpert, S.B. "DEMONSTRATION PLANTS: ARE THEY NEEDED?" (PDF). Argonne National Laboratory. Electric Power Research Institute. Retrieved 28 March 2016.

10

Industrial Engineering: Progress and Trends Since Industrial Revolution

The origin of industrial engineering can be traced back to the industrial revolution. It resulted in the expansion of the industrial sector and also gave it the modern identity which it has today. Thus, it is crucial to have an understanding of the industrial revolution for a better perspective of industrial engineering. This chapter explores some of the most significant aspects of industrial revolution and its impact on industrial engineering.

A Watt steam engine. James Watt transformed the steam engine from a reciprocating motion that was used for pumping to a rotating motion suited to industrial applications. Watt and others significantly improved the efficiency of the steam engine.

The Industrial Revolution was the transition to new manufacturing processes in the period from about 1760 to sometime between 1820 and 1840. This transition included going from hand production methods to machines, new chemical manufacturing and iron production processes, improved efficiency of water power, the increasing use of steam power, the development of machine tools and the rise of the factory system. Textiles were the dominant industry of the Industrial Revolution in terms of employment, value of output and capital invested; the textile industry was also the first to use modern production methods. The Industrial Revolution began in Great Britain and most of the important technological innovations were British.

The Industrial Revolution marks a major turning point in history; almost every aspect of daily life was influenced in some way. In particular, average income and population began to exhibit unprecedented sustained growth. Some economists say that the major impact of the Industrial Revolution was that the standard of living for the general population began to increase consistently for the first time in history, although others have said that it did not begin to meaningfully improve until the late 19th and 20th centuries. At approximately the same time the Industrial Revolution was occurring, Britain was undergoing an agricultural revolution, which also helped to improve living standards and provided surplus labour available for industry.

Mechanised textile production spread from Great Britain to continental Europe in the early 19th century, with important centres of textiles, iron and coal emerging in Belgium, and later in France. Since then industrialisation has spread throughout much of the world. The precise start and end of the Industrial Revolution is still debated among historians, as is the pace of economic and social changes. GDP per capita was broadly stable before the Industrial Revolution and the emergence of the modern capitalist economy, while the Industrial Revolution began an era of per-capita economic growth in capitalist economies. Economic historians are in agreement that the onset of the Industrial Revolution is the most important event in the history of humanity since the domestication of animals and plants.

The First Industrial Revolution evolved into the Second Industrial Revolution in the transition years between 1840 and 1870, when technological and economic progress continued with the increasing adoption of steam transport (steam-powered railways, boats and ships), the large-scale manufacture of machine tools and the increasing use of machinery in steam-powered factories.

Etymology

The earliest recorded use of the term "Industrial Revolution" seems to have been in a letter from 6 July 1799 written by French envoy Louis-Guillaume Otto, announcing that France had entered the race to industrialise. In his 1976 book *Keywords: A Vocabulary of Culture and Society*, Raymond Williams states in the entry for "Industry": "The idea of a new social order based on major industrial change was clear in Southey and Owen, between 1811 and 1818, and was implicit as early as Blake in the early 1790s and Wordsworth at the turn of the [19th] century." The term *Industrial Revolution* applied to technological change was becoming more common by the late 1830s, as in Jérôme-Adolphe Blanqui's description in 1837 of *la révolution industrielle*. Friedrich Engels in *The Condition of the Working Class in England* in 1844 spoke of "an industrial revolution, a revolution which at the same time changed the whole of civil society". However, although Engels wrote in the 1840s, his book was not translated into English until the late 1800s, and his expression did not enter everyday language until then. Credit for popularising the term may be given to Arnold Toynbee, whose 1881 lectures gave a detailed account of the term.

Some historians, such as John Clapham and Nicholas Crafts, have argued that the economic and social changes occurred gradually and the term *revolution* is a misnomer. This is still a subject of debate among historians.

Important Technological Developments

The commencement of the Industrial Revolution is closely linked to a small number of innovations, beginning in the second half of the 18th century. By the 1830s the following gains had been made in important technologies:

- Textiles – Mechanised cotton spinning powered by steam or water greatly increased the output of a worker. The power loom increased the output of a worker by a factor of over 40. The cotton gin increased productivity of removing seed from cotton by a factor of 50. Large gains in productivity also occurred in spinning and weaving of wool and linen, but they were not as great as in cotton.

- Steam power – The efficiency of steam engines increased so that they used between one-fifth and one-tenth as much fuel. The adaptation of stationary steam engines to rotary motion made them suitable for industrial uses. The high pressure engine had a high power to weight ratio, making it suitable for transportation. Steam power underwent a rapid expansion after 1800.

- Iron making – The substitution of coke for charcoal greatly lowered the fuel cost for pig iron and wrought iron production. Using coke also allowed larger blast furnaces, resulting in economies of scale. The cast iron blowing cylinder was first used in 1760. It was later improved by making it double acting, which allowed higher furnace temperatures. The puddling process produced a structural grade iron at a lower cost than the finery forge. The rolling mill was fifteen times faster than hammering wrought iron. Hot blast (1828) greatly increased fuel efficiency in iron production in the following decades.

Textile Manufacture

In the late 17th and early 18th centuries the British government passed a series of Calico Acts in order to protect the domestic woollen industry from the increasing amounts of cotton fabric imported from India.

The demand for heavier fabric was met by a domestic industry based around Lancashire that produced fustian, a cloth with flax warp and cotton weft. Flax was used for the warp because wheel spun cotton did not have sufficient strength, but the resulting blend was not as soft as 100% cotton and was more difficult to sew.

On the eve of the Industrial Revolution, spinning and weaving were done in households, for domestic consumption and as a cottage industry under the putting-out system. Occasionally the work was done in the workshop of a master weaver. Under the putting-out system, home-based workers produced under contract to merchant sellers, who often supplied the raw materials. In the off season the women, typically farmers' wives, did the spinning and the men did the weaving. Using the spinning wheel it took anywhere from four to eight spinners to supply one hand loom weaver. The flying shuttle patented in 1733 by John Kay, with a number of subsequent improvements including an important one in 1747, doubled the output of a weaver, worsening the imbalance between spinning and weaving. It became widely used around Lancashire after 1760 when John's son, Robert, invented the drop box.

patented the roller spinning frame and the flyer-and-bobbin system for drawing wool to a more even thickness. The technology was developed with the help of John Wyatt of Birmingham. Paul and Wyatt opened a mill in Birmingham which used their new rolling machine powered by a donkey. In 1743, a factory opened in Northampton with fifty spindles on each of five of Paul and

Wyatt's machines. This operated until about 1764. A similar mill was built by Daniel Bourn in Leominster, but this burnt down. Both Lewis Paul and Daniel Bourn patented carding machines in 1748. Based on two sets of rollers that travelled at different speeds, it was later used in the first cotton spinning mill. Lewis's invention was later developed and improved by Richard Arkwright in his water frame and Samuel Crompton in his spinning mule.

Model of the spinning jenny in a museum in Wuppertal. Invented by James Hargreaves in 1764, the spinning jenny was one of the innovations that started the revolution.

In 1764 in the village of Stanhill, Lancashire, James Hargreaves invented the spinning jenny, which he patented in 1770. It was the first practical spinning frame with multiple spindles. The jenny worked in a similar manner to the spinning wheel, by first clamping down on the fibres, then by drawing them out, followed by twisting. It was a simple, wooden framed machine that only cost about £6 for a 40 spindle model in 1792, and was used mainly by home spinners. The jenny produced a lightly twisted yarn only suitable for weft, not warp.

The spinning frame or water frame was developed by Richard Arkwright who, along with two partners, patented it in 1769. The design was partly based on a spinning machine built for Thomas High by clock maker John Kay, who was hired by Arkwright. For each spindle, the water frame used a series of four pairs of rollers, each operating at a successively higher rotating speed, to draw out the fibre, which was then twisted by the spindle. The roller spacing was slightly longer than the fibre length. Too close a spacing caused the fibres to break while too distant a spacing caused uneven thread. The top rollers were leather covered and loading on the rollers was applied by a weight. The weights kept the twist from backing up before the rollers. The bottom rollers were wood and metal, with fluting along the length. The water frame was able to produce a hard, medium count thread suitable for warp, finally allowing 100% cotton cloth to be made in Britain. A horse powered the first factory to use the spinning frame. Arkwright and his partners used water power at a factory in Cromford, Derbyshire in 1771, giving the invention its name.

The only surviving example of a spinning mule built by the inventor Samuel Crompton. The mule produced superior quality thread with minimal labour.

Samuel Crompton's Spinning Mule, introduced in 1779, was a combination of the spinning jenny and the water frame in which the spindles were placed on a carriage, which went through an operational sequence during which the rollers stopped while the carriage moved away from the drawing roller to finish drawing out the fibres as the spindles started rotating. Crompton's mule was able to produce finer thread than hand spinning and at a lower cost. Mule spun thread was of suitable strength to be used as warp, and finally allowed Britain to produce good quality calico cloth.

Interior of Marshall's Temple Works

Realising that the expiration of the Arkwright patent would greatly increase the supply of spun cotton and lead to a shortage of weavers, Edmund Cartwright developed a vertical power loom which he patented in 1785. In 1776 he patented a two-man operated loom, that was more conventional. Cartwright built two factories; the first burned down and the second was sabotaged by his workers. Cartwright's loom design had several flaws, the most serious being thread breakage. Samuel Horrocks patented a fairly successful loom in 1813. Horock's loom was improved by Richard Roberts in 1822 and these were produced in large numbers by Roberts, Hill & Co.

The demand for cotton presented an opportunity to planters in the Southern United States, who thought upland cotton would be a profitable crop if a better way could be found to remove the seed. Eli Whitney responded to the challenge by inventing the inexpensive cotton gin. With a cotton gin a man could remove seed from as much upland cotton in one day as would have previously taken a woman working two months to process at one pound per day.

Other inventors increased the efficiency of the individual steps of spinning (carding, twisting and spinning, and rolling) so that the supply of yarn increased greatly. This in turn fed a weaving industry that advanced with improvements to shuttles and the loom or 'frame'. The output of an individual labourer increased dramatically, with the effect that the new machines were seen as a threat to employment, and early innovators were attacked and their inventions destroyed.

To capitalise upon these advances, it took a class of entrepreneurs, of whom the best known is Richard Arkwright. He is credited with a list of inventions, but these were actually developed by such people as Thomas Highs and John Kay; Arkwright nurtured the inventors, patented the ideas, financed the initiatives, and protected the machines. He created the cotton mill which brought the production processes together in a factory, and he developed the use of power—first horse power and then water power—which made cotton manufacture a mechanised industry. Before long steam power was applied to drive textile machinery. Manchester acquired the nickname Cottonopolis during the early 19th century owing to its sprawl of textile factories.

Metallurgy

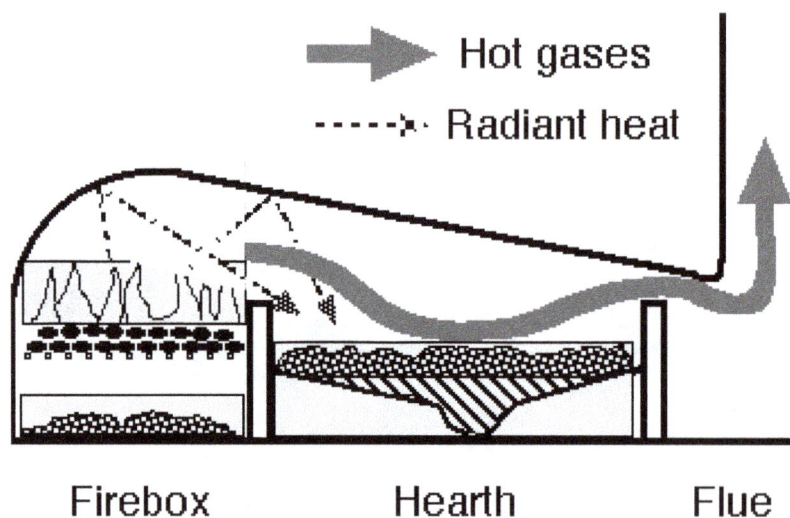

The Reverberatory Furnace could produce cast iron using mined coal. The burning coal remained separate from the iron ore and so did not contaminate the iron with impurities like sulphur and ash. This opened the way to increased iron production.

The Iron Bridge, Shropshire, England

A major change in the metal industries during the era of the Industrial Revolution was the replacement of wood and other bio-fuels with coal. For a given amount of heat, coal required much less labour to mine than cutting wood and converting it to charcoal, and coal was more abundant than wood.

Use of coal in smelting started somewhat before the Industrial Revolution, based on innovations by Sir Clement Clerke and others from 1678, using coal reverberatory furnaces known as cupolas. These were operated by the flames playing on the ore and charcoal or coke mixture, reducing the oxide to metal. This has the advantage that impurities (such as sulphur ash) in the coal do not migrate into the metal. This technology was applied to lead from 1678 and to copper from 1687. It was also applied to iron foundry work in the 1690s, but in this case the reverberatory furnace was known as an air furnace. The foundry cupola is a different (and later) innovation.

This was followed by Abraham Darby, who made great strides using coke to fuel his blast furnaces at Coalbrookdale in 1709. However, the coke pig iron he made was used mostly for the production of cast-iron goods, such as pots and kettles. He had the advantage over his rivals in that his pots, cast by his patented process, were thinner and cheaper than theirs. Coke pig iron was hardly used to produce bar iron in forges until the mid-1750s, when his son Abraham Darby II built Horsehay and Ketley furnaces (not far from Coalbrookdale). By then, coke pig iron was cheaper than charcoal pig iron. Since cast iron was becoming cheaper and more plentiful, it began being a structural material following the building of the innovative Iron Bridge in 1778 by Abraham Darby III.

Bar iron for smiths to forge into consumer goods was still made in finery forges, as it long had been. However, new processes were adopted in the ensuing years. The first is referred to today as potting and stamping, but this was superseded by Henry Cort's puddling process.

Henry Cort developed two significant iron manufacturing processes: rolling in 1783 and puddling in 1784. Rolling replaced hammering for consolidating wrought iron and expelling some of the dross. Rolling was 15 times faster than hammering with a trip hammer. Roller mills were first used for making sheets, but also were developed for rolling structural shapes such as angles and rails.

Puddling produced a structural grade iron at a relatively low cost.Puddling was a means of decarburizing pig iron by slow oxidation, with iron ore as the oxygen source, as the iron was manually stirred using a long rod. The decarburized iron, having a higher melting point than cast iron, was raked into globs by the puddler. When the glob was large enough the puddler would remove it. Puddling was backbreaking and extremely hot work. Few puddlers lived to be 40. Puddling was done in a reverberatory furnace, allowing coal or coke to be used as fuel. The puddling process continued to be used until the late 19th century when iron was being displaced by steel. Because puddling required human skill in sensing the iron globs, it was never successfully mechanised.

Up to that time, British iron manufacturers had used considerable amounts of imported iron to supplement native supplies. This came principally from Sweden from the mid-17th century and later also from Russia from the end of the 1720s. However, from 1785, imports decreased because of the new iron making technology, and Britain became an exporter of bar iron as well as manufactured wrought iron consumer goods.

Hot blast, patented by James Beaumont Neilson in 1828, was the most important development of the 19th century for saving energy in making pig iron. By using waste exhaust heat to preheat combustion air, the amount of fuel to make a unit of pig iron was reduced at first by between one-third using coal or two-thirds using coke; however, the efficiency gains continued as the technology improved. Hot blast also raised the operating temperature of furnaces, increasing their capacity. Using less coal or coke meant introducing fewer impurities into the pig iron. This meant that lower quality coal or anthracite could be used in areas where coking coal was unavailable or too expensive; however, by the end of the 19th century transportation costs fell considerably.

Two decades before the Industrial Revolution an improvement was made in the production of steel, which was an expensive commodity and used only where iron would not do, such as for cutting edge tools and for springs. Benjamin Huntsman developed his crucible steel technique in the 1740s. The raw material for this was blister steel, made by the cementation process.

The supply of cheaper iron and steel aided a number of industries, such as those making nails, hinges, wire and other hardware items. The development of machine tools allowed better working of iron, causing it to be increasingly used in the rapidly growing machinery and engine industries.

Steam Power

The development of the stationary steam engine was an important element of the Industrial Revolution; however, during the early period of the Industrial Revolution, the majority of industrial power was supplied by water and wind. In Britain by 1800 an estimated 10,000 horsepower was being supplied by steam. By 1815 steam power had grown to 210,000 hp. Small power requirements continued to be provided by animal and human muscle until the late 19th century.

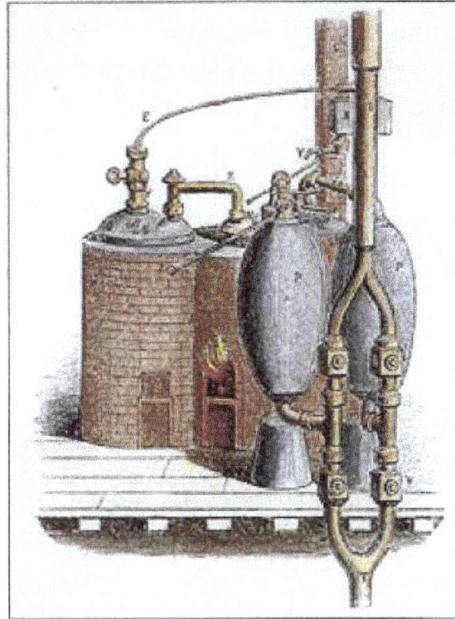

The 1698 *Savery Engine* (piston-less steam pump)– the world's first commercially useful steam powered device: built by Thomas Savery.

The first commercially successful industrial use of steam power was due to Thomas Savery in 1698. He constructed and patented in London a low-lift combined vacuum and pressure water pump, that generated about one horsepower (hp) and was used in numerous water works and in a few mines (hence its "brand name", *The Miner's Friend*). Savery's pump was economical in small horsepower ranges, but was prone to boiler explosions in larger sizes. Savery pumps continued to be produced until the late 18th century.

The first successful piston steam engine was introduced by Thomas Newcomen before 1712. A number of Newcomen engines were successfully put to use in Britain for draining hitherto unworkable deep mines, with the engine on the surface; these were large machines, requiring a lot of capital to build, and produced about 5 hp (3.7 kW). They were extremely inefficient by modern standards, but when located where coal was cheap at pit heads, opened up a great expansion in coal mining by allowing mines to go deeper. Despite their disadvantages, Newcomen engines were reliable and easy to maintain and continued to be used in the coalfields until the early decades of the 19th century. By 1729, when Newcomen died, his engines had spread (first) to Hungary in 1722, Germany, Austria, and Sweden. A total of 110 are known to have been built by 1733 when the joint patent expired, of which 14 were abroad. In the 1770s, the engineer John Smeaton built some very large examples and introduced a number of improvements. A total of 1,454 engines had been built by 1800.

A fundamental change in working principles was brought about by Scotsman James Watt. In close collaboration with Englishman Matthew Boulton, he had succeeded by 1778 in perfecting his steam engine, which incorporated a series of radical improvements, notably the closing off of the upper part of the cylinder thereby making the low pressure steam drive the top of the piston instead of the atmosphere, use of a steam jacket and the celebrated separate steam condenser chamber. The separate condenser did away with the cooling water that had been injected directly into the cylinder, which cooled the cylinder and wasted steam. Likewise, the steam jacket kept steam from condensing in the cylinder, also improving efficiency. These improvements increased engine efficiency so that Boulton & Watts engines used only 20–25% as much coal per horsepower-hour as Newcomen's. Boulton and Watt opened the Soho Foundry, for the manufacture of such engines, in 1795.

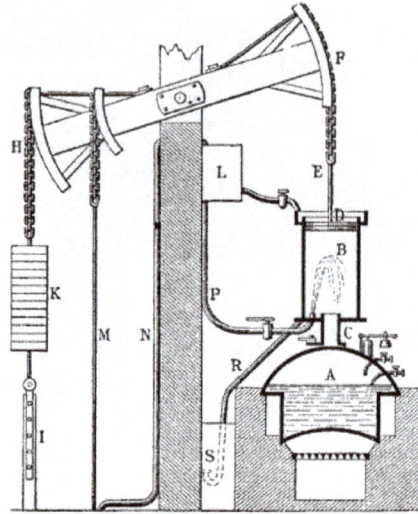

Newcomen's steam powered atmospheric engine was the first practical piston steam engine. Subsequent steam engines were to power the Industrial Revolution.

By 1783 the Watt steam engine had been fully developed into a double-acting rotative type, which meant that it could be used to directly drive the rotary machinery of a factory or mill. Both of Watt's basic engine types were commercially very successful, and by 1800, the firm Boulton & Watt had constructed 496 engines, with 164 driving reciprocating pumps, 24 serving blast furnaces, and 308 powering mill machinery; most of the engines generated from 5 to 10 hp (7.5 kW).

The development of machine tools, such as the engine lathe, planing, milling and shaping machines powered by these engines, enabled all the metal parts of the engines to be easily and accurately cut and in turn made it possible to build larger and more powerful engines.

Until about 1800, the most common pattern of steam engine was the beam engine, built as an integral part of a stone or brick engine-house, but soon various patterns of self-contained rotative engines (readily removable, but not on wheels) were developed, such as the table engine. Around the start of the 19th century, the Cornish engineer Richard Trevithick, and the American, Oliver Evans began to construct higher pressure non-condensing steam engines, exhausting against the atmosphere. High pressure yielded an engine and boiler compact enough to be used on mobile road and rail locomotives and steam boats.

Machine Tools

The Industrial Revolution created a demand for metal parts used in machinery. This led to the development of several machine tools for cutting metal parts. They have their origins in the tools developed in the 18th century by makers of clocks and watches and scientific instrument makers to enable them to batch-produce small mechanisms.

FIGURE 15. MAUDSLAY'S SCREW-CUTTING LATHE
ABOUT 1797

FIGURE 16. MAUDSLAY'S SCREW-CUTTING LATHE
ABOUT 1800

Maudslay's famous early screw-cutting lathes of circa 1797 and 1800

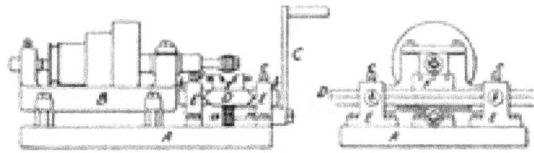

The Middletown milling machine of circa 1818, associated with Robert Johnson and Simeon North

Before the advent of machine tools, metal was worked manually using the basic hand tools of hammers, files, scrapers, saws and chisels. Consequently, the use of metal was kept to a minimum. Wooden components had the disadvantage of changing dimensions with temperature and humidity, and the various joints tended to rack (work loose) over time. As the Industrial Revolution progressed, machines with metal parts and frames became more common. Hand methods of production were very laborious and costly and precision was difficult to achieve. Pre-industrial machinery was built by various craftsmen—millwrights built water and wind mills, carpenters made wooden framing, and smiths and turners made metal parts.

The first large machine tool was the cylinder boring machine used for boring the large-diameter cylinders on early steam engines. The planing machine, the milling machine and the shaping machine were developed in the early decades of the 19th century. Although the milling machine was invented at this time, it was not developed as a serious workshop tool until somewhat later in the 19th century.

Henry Maudslay, who trained a school of machine tool makers early in the 19th century, was a mechanic with superior ability who had been employed at the Royal Arsenal, Woolwich. He was hired away by Joseph Bramah for the production of high security metal locks that required precision craftsmanship. Bramah patented a lathe that had similarities to the slide rest lathe.

Maudslay perfected the slide rest lathe, which could cut machine screws of different thread pitches by using changeable gears between the spindle and the lead screw. Before its invention screws could not be cut to any precision using various earlier lathe designs, some of which copied from a template. The slide rest lathe was called one of history's most important inventions, although not entirely Maudslay's idea.

Maudslay left Bramah's employment and set up his own shop. He was engaged to build the machinery for making ships' pulley blocks for the Royal Navy in the Portsmouth Block Mills. These machines were all-metal and were the first machines for mass production and making components with a degree of interchangeability. The lessons Maudslay learned about the need for stability and precision he adapted to the development of machine tools, and in his workshops he trained a generation of men to build on his work, such as Richard Roberts, Joseph Clement and Joseph Whitworth.

James Fox of Derby had a healthy export trade in machine tools for the first third of the century, as did Matthew Murray of Leeds. Roberts was a maker of high-quality machine tools and a pioneer of the use of jigs and gauges for precision workshop measurement.

The impact of machine tools during the Industrial Revolution was not that great because other than firearms, threaded fasteners and a few other industries there were few mass-produced metal parts. In the half century following the invention of the fundamental machine tools the machine industry became the largest industrial sector of the economy, by value added, in the U.S.

Chemicals

The large scale production of chemicals was an important development during the Industrial Revolution. The first of these was the production of sulphuric acid by the lead chamber process invented by the Englishman John Roebuck (James Watt's first partner) in 1746. He was able to greatly increase the scale of the manufacture by replacing the relatively expensive glass vessels formerly used with larger, less expensive chambers made of riveted sheets of lead. Instead of making a small amount each time, he was able to make around 100 pounds (50 kg) in each of the chambers, at least a tenfold increase.

The production of an alkali on a large scale became an important goal as well, and Nicolas Leblanc succeeded in 1791 in introducing a method for the production of sodium carbonate. The Leblanc process was a reaction of sulphuric acid with sodium chloride to give sodium sulphate and hydrochloric acid. The sodium sulphate was heated with limestone (calcium carbonate) and coal to give a mixture of sodium carbonate and calcium sulphide. Adding water separated the soluble sodium carbonate from the calcium sulphide. The process produced a large amount of pollution (the hydrochloric acid was initially vented to the air, and calcium sulphide was a useless waste product). Nonetheless, this synthetic soda ash proved economical compared to that from burning specific plants (barilla) or from kelp, which were the previously dominant sources of soda ash, and also to potash (potassium carbonate) derived from hardwood ashes.

These two chemicals were very important because they enabled the introduction of a host of other inventions, replacing many small-scale operations with more cost-effective and controllable processes. Sodium carbonate had many uses in the glass, textile, soap, and paper industries. Early uses for sulphuric acid included pickling (removing rust) iron and steel, and for bleaching cloth.

The development of bleaching powder (calcium hypochlorite) by Scottish chemist Charles Tennant in about 1800, based on the discoveries of French chemist Claude Louis Berthollet, revolutionised the bleaching processes in the textile industry by dramatically reducing the time required (from months to days) for the traditional process then in use, which required repeated exposure to the sun in bleach fields after soaking the textiles with alkali or sour milk. Tennant's factory at St Rollox, North Glasgow, became the largest chemical plant in the world.

After 1860 the focus on chemical innovation was in dyestuffs, and Germany took world leadership, building a strong chemical industry. Aspiring chemists flocked to German universities in the 1860–1914 era to learn the latest techniques. British scientists by contrast, lacked research universities and did not train advanced students; instead the practice was to hire German-trained chemists.

Cement

The Thames Tunnel (opened 1843).
Cement was used in the world's first underwater tunnel.

In 1824 Joseph Aspdin, a British bricklayer turned builder, patented a chemical process for making portland cement which was an important advance in the building trades. This process involves sintering a mixture of clay and limestone to about 1,400 °C (2,552 °F), then grinding it into a fine powder which is then mixed with water, sand and gravel to produce concrete. Portland cement was used by the famous English engineer Marc Isambard Brunel several years later when constructing the Thames Tunnel. Cement was used on a large scale in the construction of the London sewerage system a generation later.

Gas Lighting

Another major industry of the later Industrial Revolution was gas lighting. Though others made a similar innovation elsewhere, the large-scale introduction of this was the work of William Murdoch, an employee of Boulton and Watt, the Birmingham steam engine pioneers. The process consisted of the large-scale gasification of coal in furnaces, the purification of the gas (removal of sulphur, ammonia, and heavy hydrocarbons), and its storage and distribution. The first gas lighting utilities were established in London between 1812 and 1820. They soon became one of the major consumers of coal in the UK. Gas lighting affected social and industrial organisation because it allowed factories and stores to remain open longer than with tallow candles or oil. Its introduction allowed night life to flourish in cities and towns as interiors and streets could be lighted on a larger scale than before.

Glass Making

The Crystal Palace held the Great Exhibition of 1851

A new method of producing glass, known as the cylinder process, was developed in Europe during the early 19th century. In 1832, this process was used by the Chance Brothers to create sheet glass. They became the leading producers of window and plate glass. This advancement allowed for larger panes of glass to be created without interruption, thus freeing up the space planning in interiors as well as the fenestration of buildings. The Crystal Palace is the supreme example of the use of sheet glass in a new and innovative structure.

Paper Machine

A machine for making a continuous sheet of paper on a loop of wire fabric was patented in 1798 by Nicholas Louis Robert who worked for Saint-Léger Didot family in France. The paper machine is known as a Fourdrinier after the financiers, brothers Sealy and Henry Fourdrinier, who were stationers in London. Although greatly improved and with many variations, the Fourdriner machine is the predominant means of paper production today.

The method of continuous production demonstrated by the paper machine influenced the development of continuous rolling of iron and later steel and other continuous production processes.

Agriculture

The British Agricultural Revolution is considered one of the causes of the Industrial Revolution because improved agricultural productivity freed up workers to work in other sectors of the economy.

Industrial technologies that affected farming included the seed drill, the Dutch plough, which contained iron parts, and the threshing machine.

Jethro Tull invented an improved seed drill in 1701. It was a mechanical seeder which distributed seeds evenly across a plot of land and planted them at the correct depth. This was important because the yield of seeds harvested to seeds planted at that time was around four or five. Tull's seed drill was very expensive and not very reliable and therefore did not have much of an impact. Good quality seed drills were not produced until the mid 18th century.

Joseph Foljambe's *Rotherham plough* of 1730, was the first commercially successful iron plough. The threshing machine, invented by Andrew Meikle in 1784, displaced hand threshing with a flail, a laborious job that took about one-quarter of agricultural labour. It took several decades to diffuse and was the final straw for many farm labourers, who faced near starvation, leading to the 1830 agricultural rebellion of the Swing Riots.

Machine tools and metalworking techniques developed during the Industrial Revolution eventually resulted in precision manufacturing techniques in the late 19th century for mass-producing agricultural equipment, such as reapers, binders and combine harvesters.

Mining

Coal mining in Britain, particularly in South Wales started early. Before the steam engine, pits were often shallow bell pits following a seam of coal along the surface, which were abandoned as the coal was extracted. In other cases, if the geology was favourable, the coal was mined by means of an adit or drift mine driven into the side of a hill. Shaft mining was done in some areas, but the limiting factor was the problem of removing water. It could be done by hauling buckets of water up the shaft or to a sough (a tunnel driven into a hill to drain a mine). In either case, the water had to be discharged into a stream or ditch at a level where it could flow away by gravity. The introduction of the steam pump by Savery in 1698 and the Newcomen steam engine in 1712 greatly facilitated the removal of water and enabled shafts to be made deeper, enabling more coal to be extracted. These were developments that had begun before the Industrial Revolution, but the adoption of John Smeaton's improvements to the Newcomen engine followed by James Watt's more efficient steam engines from the 1770s reduced the fuel costs of engines, making mines more profitable.

Coal mining was very dangerous owing to the presence of firedamp in many coal seams. Some degree of safety was provided by the safety lamp which was invented in 1816 by Sir Humphry Davy and independently by George Stephenson. However, the lamps proved a false dawn because they became unsafe very quickly and provided a weak light. Firedamp explosions continued, often setting off coal dust explosions, so casualties grew during the entire 19th century. Conditions of work were very poor, with a high casualty rate from rock falls.

Other Developments

Other developments included more efficient water wheels, based on experiments conducted by the British engineer John Smeaton the beginnings of a machine industry and the rediscovery of concrete (based on hydraulic lime mortar) by John Smeaton, which had been lost for 1300 years.

Transportation

At the beginning of the Industrial Revolution, inland transport was by navigable rivers and roads, with coastal vessels employed to move heavy goods by sea. Wagon ways were used for conveying coal to rivers for further shipment, but canals had not yet been widely constructed. Animals supplied all of the motive power on land, with sails providing the motive power on the sea. The first horse railways were introduced toward the end of the 18th century, with steam locomotives being introduced in the early decades of the 19th century.

The Industrial Revolution improved Britain's transport infrastructure with a turnpike road network, a canal and waterway network, and a railway network. Raw materials and finished products could be moved more quickly and cheaply than before. Improved transportation also allowed new ideas to spread quickly.

Canals

The Bridgewater Canal, famous because of its commercial success, crossing the Manchester Ship Canal, one of the last canals to be built.

Canals were the first technology to allow bulk materials to be economically transported long distances inland. This was because a horse could pull a barge with a load dozens of times larger than the load that could be drawn in a cart.

Building of canals dates to ancient times. The Grand Canal in China, "the world's largest artificial waterway and oldest canal still in existence," parts of which were started between the 6th and 4th centuries BC, is 1,121 miles (1,804 km) long and links Hangzhou with Beijing.

In the UK, canals began to be built in the late 18th century to link the major manufacturing centres across the country. Known for its huge commercial success, the Bridgewater Canal in North West England, which opened in 1761 and was mostly funded by The 3rd Duke of Bridgewater. From Worsley to the rapidly growing town of Manchester its construction cost £168,000 (£22,589,130 as of 2013), but its advantages over land and river transport meant that within a year of its opening in 1761, the price of coal in Manchester fell by about half. This success helped inspire a period of intense canal building, known as Canal Mania. New canals were hastily built in the aim of replicating the commercial success of the Bridgewater Canal, the most notable being the Leeds and Liverpool Canal and the Thames and Severn Canal which opened in 1774 and 1789 respectively.

By the 1820s, a national network was in existence. Canal construction served as a model for the organisation and methods later used to construct the railways. They were eventually largely superseded as profitable commercial enterprises by the spread of the railways from the 1840s on. The last major canal to be built in the United Kingdom was the Manchester Ship Canal, which upon opening in 1894 was the largest ship canal in the world, and opened Manchester as a port. However it never achieved the commercial success its sponsors had hoped for and signalled canals as a dying mode of transport in an age dominated by railways, which were quicker and often cheaper.

Britain's canal network, together with its surviving mill buildings, is one of the most enduring features of the early Industrial Revolution to be seen in Britain.

Roads

Construction of the first macadam road in the United States (1823). In the foreground, workers are breaking stones "so as not to exceed 6 ounces in weight or to pass a two-inch ring".

Much of the original British road system was poorly maintained by thousands of local parishes, but from the 1720s (and occasionally earlier) turnpike trusts were set up to charge tolls and maintain some roads. Increasing numbers of main roads were turnpiked from the 1750s to the extent that almost every main road in England and Wales was the responsibility of a turnpike trust. New engineered roads were built by John Metcalf, Thomas Telford and most notably John McAdam, with the first 'macadamised' stretch of road being Marsh Road at Ashton Gate, Bristol in 1816. The major turnpikes radiated from London and were the means by which the Royal Mail was able to reach the rest of the country. Heavy goods transport on these roads was by means of slow, broad wheeled, carts hauled by teams of horses. Lighter goods were conveyed by smaller carts or by teams of pack horse. Stage coaches carried the rich, and the less wealthy could pay to ride on carriers carts.

Railways

Painting depicting the opening of the Liverpool and Manchester Railway in 1830, the first inter-city railway in the world and which spawned Railway Mania due to its success.

Reducing friction was one of the major reasons for the success of railroads compared to wagons. This was demonstrated on an iron plate covered wooden tramway in 1805 at Croydon, England.

" A good horse on an ordinary turnpike road can draw two thousand pounds, or one ton. A party of gentlemen were invited to witness the experiment, that the superiority of the new road might be established by ocular demonstration. Twelve wagons were loaded with stones, till each wagon weighed three tons, and the wagons were fastened together. A horse was then attached, which drew the wagons with ease, six miles in two hours, having stopped four times, in order to show he had the power of starting, as well as drawing his great load."

Railways were made practical by the widespread introduction of inexpensive puddled iron after 1800, the rolling mill for making rails, and the development of the high pressure steam engine also around 1800.

Wagonways for moving coal in the mining areas had started in the 17th century and were often associated with canal or river systems for the further movement of coal. These were all horse drawn or relied on gravity, with a stationary steam engine to haul the wagons back to the top of the incline. The first applications of the steam locomotive were on wagon or plate ways (as they were then often called from the cast-iron plates used). Horse-drawn public railways did not begin until the early years of the 19th century when improvements to pig and wrought iron production were lowering costs.

Steam locomotives began being built after the introduction of high pressure steam engines after the expiration of the Boulton and Watt patent in 1800. High pressure engines exhausted used steam to the atmosphere, doing away with the condenser and cooling water. They were also much lighter weight and smaller in size for a given horsepower than the stationary condensing engines.

A few of these early locomotives were used in mines. Steam-hauled public railways began with the Stockton and Darlington Railway in 1825.

The rapid introduction of railways followed the 1829 Rainhill Trials, which demonstrated Robert Stephenson's successful locomotive design and the 1828 development of Hot blast, which dramatically reduced the fuel consumption of making iron and increased the capacity the blast furnace.

On 15 September 1830, the Liverpool and Manchester Railway was opened, the first inter-city railway in the world and was attended by Prime Minister, the Duke of Wellington. The railway was engineered by Joseph Locke and George Stephenson, linked the rapidly expanding industrial town of Manchester with the port town of Liverpool. The opening was marred by problems, due to the primitive nature of the technology being employed, however problems were gradually ironed out and the railway became highly successful, transporting passengers and freight. The success of the inter-city railway, particularly in the transport of freight and commodities, led to Railway Mania.

Construction of major railways connecting the larger cities and towns began in the 1830s but only gained momentum at the very end of the first Industrial Revolution. After many of the workers had completed the railways, they did not return to their rural lifestyles but instead remained in the cities, providing additional workers for the factories.

Social Effects

Prior to the Industrial Revolution most of the workforce was employed in agriculture, either as self-employed farmers as land owners or tenants, or as landless agricultural labourers. By the time of the Industrial Revolution the putting-out system whereby farmers and townspeople produced goods in their homes, often described as *cottage industry*, was the standard. Typical putting out system goods included spinning and weaving. Merchant capitalist provided the raw materials, typically paid workers by the piece, and were responsible for the sale of the goods. Embezzlement of supplies by workers and poor quality were common problems. The logistical effort in procuring and distributing raw materials and picking up finished goods were also limitations of the putting out system.

Some early spinning and weaving machinery, such as a 40 spindle jenny for about 6 pounds in 1792, was affordable for cottagers. Later machinery such as spinning frames, spinning mules and power looms were expensive (especially if water powered), giving rise to capitalist ownership of factories. Many workers, who had nothing but their labour to sell, became factory workers out of necessity.

The change in the social relationship of the factory worker compared to farmers and cottagers was viewed unfavourably by Karl Marx, however, he recognized the increase in productivity made possible by technology.

Impact on Women and Family Life

Women's historians have debated the effect of the Industrial Revolution and capitalism generally on the status of women. Taking a pessimistic side, Alice Clark argued that when capitalism arrived in 17th century England, it lowered the status of women as they lost much of their economic importance. Clark argues that in 16th century England, women were engaged in many aspects of industry and agriculture. The home was a central unit of production and women played a vital role in running farms, and in some trades and landed estates. Their useful economic roles gave them a sort of equality with their husbands. However, Clark argues, as capitalism expanded in the 17th century, there was more and more division of labour with the husband taking paid labour jobs outside the home, and the wife reduced to unpaid household work. Middle-class and women were confined to an idle domestic existence, supervising servants; lower-class women were forced to take poorly paid jobs. Capitalism, therefore, had a negative effect on powerful women.

In a more positive interpretation, Ivy Pinchbeck argues that capitalism created the conditions for women's emancipation. Tilly and Scott have emphasised the continuity in the status of women, finding three stages in English history. In the pre-industrial era, production was mostly for home use and women produce much of the needs of the households. The second stage was the "family wage economy" of early industrialisation; the entire family depended on the collective wages of its members, including husband, wife and older children. The third or modern stage is the "family consumer economy," in which the family is the site of consumption, and women are employed in large numbers in retail and clerical jobs to support rising standards of consumption.

Standards of Living

The effects on living conditions the industrial revolution have been very controversial, and were hotly debated by economic and social historians from the 1950s to the 1980s. A series of 1950s essays by Henry Phelps Brown and Sheila V. Hopkins later set the academic consensus that the bulk of the population, that was at the bottom of the social ladder, suffered severe reductions in their living standards. During 1813–1913, there was a significant increase in worker wages.

Some economists, such as Robert E. Lucas, Jr., say that the real impact of the Industrial Revolution was that "for the first time in history, the living standards of the masses of ordinary people have begun to undergo sustained growth ... Nothing remotely like this economic behaviour is mentioned by the classical economists, even as a theoretical possibility." Others, however, argue that while growth of the economy's overall productive powers was unprecedented during the Industrial Revolution, living standards for the majority of the population did not grow meaningfully until the late 19th and 20th centuries, and that in many ways workers' living standards declined under early capitalism: for instance, studies have shown that real wages in Britain only increased 15% between the 1780s and 1850s, and that life expectancy in Britain did not begin to dramatically increase until the 1870s.

Food and Nutrition

Chronic hunger and malnutrition were the norm for the majority of the population of the world including Britain and France, until the late 19th century. Until about 1750, in large part due to malnutrition, life expectancy in France was about 35 years, and only slightly higher in Britain. The United States population of the time was adequately fed, much taller on average and had life expectancy of 45–50 years.

In Britain and the Netherlands, food supply had been increasing and prices falling before the Industrial Revolution due to better agricultural practices; however, population grew too, as noted by Thomas Malthus.> Before the Industrial Revolution, advances in agriculture or technology soon led to an increase in population, which again strained food and other resources, limiting increases in per capita income. This condition is called the Malthusian trap, and it was finally overcome by industrialisation.

Transportation improvements, such as canals and improved roads, also lowered food costs. Railroads were introduced near the end of the Industrial Revolution.

Housing

Living conditions during the Industrial Revolution varied from splendour for factory owners to squalor for workers.

In *The Condition of the Working Class in England* in 1844 Friedrich Engels described backstreet sections of Manchester and other mill towns, where people lived in crude shanties and shacks, some not completely enclosed, some with dirt floors. These shanty towns had narrow walkways between irregularly shaped lots and dwellings. There were no sanitary facilities. Population density was extremely high. Eight to ten unrelated mill workers often shared a room, often with no furniture, and slept on a pile of straw or sawdust. Toilet facilities were shared if they existed. Disease spread through a contaminated water supply. Also, people were at risk of developing pathologies due to persistent dampness.

The famines that troubled rural areas did not happen in industrial areas. But urban people—especially small children—died due to diseases spreading through the cramped living conditions. Tuberculosis (spread in congested dwellings), lung diseases from the mines, cholera from polluted water and typhoid were also common.

Not everyone lived in such poor conditions. The Industrial Revolution also created a middle class of professionals, such as lawyers and doctors, who lived in much better conditions.

Conditions improved over the course of the 19th century due to new public health acts regulating things such as sewage, hygiene and home construction. In the introduction of his 1892 edition, Engels notes that most of the conditions he wrote about in 1844 had been greatly improved.

Clothing and Consumer Goods

Consumers benefited from falling prices for clothing and household articles such as cast iron cooking utensils, and in the following decades, stoves for cooking and space heating.

Population Increase

According to Robert Hughes in *The Fatal Shore*, the population of England and Wales, which had remained steady at 6 million from 1700 to 1740, rose dramatically after 1740. The population of England had more than doubled from 8.3 million in 1801 to 16.8 million in 1850 and, by 1901, had nearly doubled again to 30.5 million. Improved conditions led to the population of Britain increasing from 10 million to 40 million in the 1800s. Europe's population increased from about 100 million in 1700 to 400 million by 1900.

The Industrial Revolution was the first period in history during which there was a simultaneous increase in population and in per capita income.

Labour Conditions

Social Structure and Working Conditions

In terms of social structure, the Industrial Revolution witnessed the triumph of a middle class of industrialists and businessmen over a landed class of nobility and gentry. Ordinary working people found increased opportunities for employment in the new mills and factories, but these were often under strict working conditions with long hours of labour dominated by a pace set by machines. As late as the year 1900, most industrial workers in the United States still worked a 10-hour day (12 hours in the steel industry), yet earned from 20% to 40% less than the minimum deemed necessary for a decent life. However, harsh working conditions were prevalent long before the Industrial Revolution took place. Pre-industrial society was very static and often cruel—child labour, dirty living conditions, and long working hours were just as prevalent before the Industrial Revolution.

Factories and Urbanisation

Manchester, England ("Cottonopolis"), pictured in 1840, showing the mass of factory chimneys

Industrialisation led to the creation of the factory. Arguably the first highly mechanised was John Lombe's water-powered silk mill at Derby, operational by 1721. Lombe learned silk thread manufacturing by taking a job in Italy and acting as an industrial spy; however, since the silk industry there was a closely guarded secret, the state of the industry there is unknown. Because Lombe's factory was not successful and there was no follow through, the rise of the modern factory dates to somewhat later when cotton spinning was mechanised.

The factory system contributed to the growth of urban areas, as large numbers of workers migrated into the cities in search of work in the factories. Nowhere was this better illustrated than the mills and associated industries of Manchester, nicknamed "Cottonopolis", and the world's first industrial city. Manchester experienced a six-times increase in its population between 1771 and 1831. Bradford grew by 50% every ten years between 1811 and 1851 and by 1851 only 50% of the population of Bradford was actually born there.

For much of the 19th century, production was done in small mills, which were typically water-powered and built to serve local needs. Later, each factory would have its own steam engine and a chimney to give an efficient draft through its boiler.

The transition to industrialisation was not without difficulty. For example, a group of English workers known as Luddites formed to protest against industrialisation and sometimes sabotaged factories.

In other industries the transition to factory production was not so divisive. Some industrialists themselves tried to improve factory and living conditions for their workers. One of the earliest such reformers was Robert Owen, known for his pioneering efforts in improving conditions for workers at the New Lanark mills, and often regarded as one of the key thinkers of the early socialist movement.

By 1746, an integrated brass mill was working at Warmley near Bristol. Raw material went in at one end, was smelted into brass and was turned into pans, pins, wire, and other goods. Housing was provided for workers on site. Josiah Wedgwood and Matthew Boulton (whose Soho Manufactory was completed in 1766) were other prominent early industrialists, who employed the factory system.

Child Labour

A young "drawer" pulling a coal tub along a mine gallery. In Britain laws passed
in 1842 and 1844 improved mine working conditions.

The Industrial Revolution led to a population increase but the chances of surviving childhood did not improve throughout the Industrial Revolution, although *infant* mortality rates were reduced markedly. There was still limited opportunity for education and children were expected to work. Employers could pay a child less than an adult even though their productivity was comparable; there was no need for strength to operate an industrial machine, and since the industrial system was completely new, there were no experienced adult labourers. This made child labour the labour of choice for manufacturing in the early phases of the Industrial Revolution between the

18th and 19th centuries. In England and Scotland in 1788, two-thirds of the workers in 143 water-powered cotton mills were described as children.

Child labour existed before the Industrial Revolution but with the increase in population and education it became more visible. Many children were forced to work in relatively bad conditions for much lower pay than their elders, 10–20% of an adult male's wage. Children as young as four were employed. Beatings and long hours were common, with some child coal miners and hurriers working from 4 am until 5 pm. Conditions were dangerous, with some children killed when they dozed off and fell into the path of the carts, while others died from gas explosions. Many children developed lung cancer and other diseases and died before the age of 25. Workhouses would sell orphans and abandoned children as "pauper apprentices", working without wages for board and lodging. Those who ran away would be whipped and returned to their masters, with some masters shackling them to prevent escape. Children employed as mule scavengers by cotton mills would crawl under machinery to pick up cotton, working 14 hours a day, six days a week. Some lost hands or limbs, others were crushed under the machines, and some were decapitated. Young girls worked at match factories, where phosphorus fumes would cause many to develop phossy jaw. Children employed at glassworks were regularly burned and blinded, and those working at potteries were vulnerable to poisonous clay dust.

Reports were written detailing some of the abuses, particularly in the coal mines and textile factories, and these helped to popularise the children's plight. The public outcry, especially among the upper and middle classes, helped stir change in the young workers' welfare.

Politicians and the government tried to limit child labour by law but factory owners resisted; some felt that they were aiding the poor by giving their children money to buy food to avoid starvation, and others simply welcomed the cheap labour. In 1833 and 1844, the first general laws against child labour, the Factory Acts, were passed in Britain: Children younger than nine were not allowed to work, children were not permitted to work at night, and the work day of youth under the age of 18 was limited to twelve hours. Factory inspectors supervised the execution of the law, however, their scarcity made enforcement difficult. About ten years later, the employment of children and women in mining was forbidden. These laws decreased the number of child labourers, however child labour remained in Europe and the United States up to the 20th century.

Luddites

The rapid industrialisation of the English economy cost many craft workers their jobs. The movement started first with lace and hosiery workers near Nottingham and spread to other areas of the textile industry owing to early industrialisation. Many weavers also found themselves suddenly unemployed since they could no longer compete with machines which only required relatively limited (and unskilled) labour to produce more cloth than a single weaver. Many such unemployed workers, weavers and others, turned their animosity towards the machines that had taken their jobs and began destroying factories and machinery. These attackers became known as Luddites, supposedly followers of Ned Ludd, a folklore figure. The first attacks of the Luddite movement began in 1811. The Luddites rapidly gained popularity, and the British government took drastic measures, using the militia or army to protect industry. Those rioters who were caught were tried and hanged, or transported for life.

Luddites smashing a power loom in 1812

Unrest continued in other sectors as they industrialised, such as with agricultural labourers in the 1830s when large parts of southern Britain were affected by the Captain Swing disturbances. Threshing machines were a particular target, and hayrick burning was a popular activity. However, the riots led to the first formation of trade unions, and further pressure for reform.

Organisation of Labour

The Industrial Revolution concentrated labour into mills, factories and mines, thus facilitating the organisation of *combinations* or trade unions to help advance the interests of working people. The power of a union could demand better terms by withdrawing all labour and causing a consequent cessation of production. Employers had to decide between giving in to the union demands at a cost to themselves or suffering the cost of the lost production. Skilled workers were hard to replace, and these were the first groups to successfully advance their conditions through this kind of bargaining.

The main method the unions used to effect change was strike action. Many strikes were painful events for both sides, the unions and the management. In Britain, the Combination Act 1799 forbade workers to form any kind of trade union until its repeal in 1824. Even after this, unions were still severely restricted.

In 1832, the Reform Act extended the vote in Britain but did not grant universal suffrage. That year six men from Tolpuddle in Dorset founded the Friendly Society of Agricultural Labourers to protest against the gradual lowering of wages in the 1830s. They refused to work for less than ten shillings a week, although by this time wages had been reduced to seven shillings a week and were due to be further reduced to six. In 1834 James Frampton, a local landowner, wrote to the Prime

Minister, Lord Melbourne, to complain about the union, invoking an obscure law from 1797 prohibiting people from swearing oaths to each other, which the members of the Friendly Society had done. James Brine, James Hammett, George Loveless, George's brother James Loveless, George's brother in-law Thomas Standfield, and Thomas's son John Standfield were arrested, found guilty, and transported to Australia. They became known as the Tolpuddle Martyrs. In the 1830s and 1840s, the Chartist movement was the first large-scale organised working class political movement which campaigned for political equality and social justice. Its *Charter* of reforms received over three million signatures but was rejected by Parliament without consideration.

Working people also formed friendly societies and co-operative societies as mutual support groups against times of economic hardship. Enlightened industrialists, such as Robert Owen also supported these organisations to improve the conditions of the working class.

Unions slowly overcame the legal restrictions on the right to strike. In 1842, a general strike involving cotton workers and colliers was organised through the Chartist movement which stopped production across Great Britain.

Eventually, effective political organisation for working people was achieved through the trades unions who, after the extensions of the franchise in 1867 and 1885, began to support socialist political parties that later merged to become the British Labour Party.

Impact on Environment

Levels of air pollution rose during the Industrial Revolution, sparking the first modern environmental laws to be passed in the mid-19th century.

The origins of the environmental movement lay in the response to increasing levels of smoke pollution in the atmosphere during the Industrial Revolution. The emergence of great factories and the concomitant immense growth in coal consumption gave rise to an unprecedented level of air pollution in industrial centers; after 1900 the large volume of industrial chemical discharges added to the growing load of untreated human waste. The first large-scale, modern environmental laws came in the form of Britain's Alkali Acts, passed in 1863, to regulate the deleterious air pollution (gaseous hydrochloric acid) given off by the Leblanc process, used to produce soda ash. An Alkali inspector and four sub-inspectors were appointed to curb this pollution. The responsibilities of the inspectorate were gradually expanded, culminating in the Alkali Order 1958 which placed all major heavy industries that emitted smoke, grit, dust and fumes under supervision.

The manufactured gas industry began in British cities in 1812-1820. The technique used produced highly toxic effluent that was dumped into sewers and rivers. The gas companies were repeatedly sued in nuisance lawsuits. They usually lost and modified the worst practices. The City of London repeatedly indicted gas companies in the 1820s for polluting the Thames and poisoning its fish. Finally, Parliament wrote company charters to regulate toxicity. The industry reached the US around 1850 causing pollution and lawsuits.

In industrial cities local experts and reformers, especially after 1890, took the lead in identifying environmental degradation and pollution, and initiating grass-roots movements to demand and achieve reforms. Typically the highest priotiry went to water and air pollution. The Coal Smoke Abatement Society was formed in Britain in 1898 making it one of the oldest environmental NGOs. It was founded by artist Sir William Blake Richmond, frustrated with the pall cast by coal smoke. Although there were earlier pieces of legislation, the Public Health Act 1875 required all furnaces and fireplaces to consume their own smoke. It also provided for sanctions against factories that emitted large amounts of black smoke. The provisions of this law were extended in 1926 with the Smoke Abatement Act to include other emissions, such as soot, ash and gritty particles and to empower local authorities to impose their own regulations.

Other Effects

The application of steam power to the industrial processes of printing supported a massive expansion of newspaper and popular book publishing, which reinforced rising literacy and demands for mass political participation.

During the Industrial Revolution, the life expectancy of children increased dramatically. The percentage of the children born in London who died before the age of five decreased from 74.5% in 1730–1749 to 31.8% in 1810–1829.

The growth of modern industry since the late 18th century led to massive urbanisation and the rise of new great cities, first in Europe and then in other regions, as new opportunities brought huge numbers of migrants from rural communities into urban areas. In 1800, only 3% of the world's population lived in cities, compared to nearly 50% today (the beginning of the 21st century). Manchester had a population of 10,000 in 1717, but by 1911 it had burgeoned to 2.3 million.

Industrialisation Beyond the United Kingdom

Continental Europe

Eric Hobsbawm held that the Industrial Revolution began in Britain in the 1780s and was not fully felt until the 1830s or 1840s, while T. S. Ashton held that it occurred roughly between 1760 and 1830. The Industrial Revolution on Continental Europe came a little later than in Great Britain. In many industries, this involved the application of technology developed in Britain in new places. Often the technology was purchased from Britain or British engineers and entrepreneurs moved abroad in search of new opportunities. By 1809, part of the Ruhr Valley in Westphalia was called 'Miniature England' because of its similarities to the industrial areas of England. The German, Russian and Belgian governments all provided state funding to the new industries. In some cases (such as iron), the different availability of resources locally meant that only some aspects of the British technology were adopted.

Belgium

Belgium was the second country, after Britain, in which the Industrial Revolution took place and the first in continental Europe: Wallonia (French speaking southern Belgium) was the first region to follow the British model successfully. Starting in the middle of the 1820s, and especially after Belgium became an independent nation in 1830, numerous works comprising coke blast furnaces as well as puddling and rolling mills were built in the coal mining areas around Liège and Charleroi. The leader was a transplanted Englishman John Cockerill. His factories at Seraing integrated all stages of production, from engineering to the supply of raw materials, as early as 1825.

Wallonia exemplified the radical evolution of industrial expansion. Thanks to coal (the French word "houille" was coined in Wallonia), the region geared up to become the 2nd industrial power in the world after Britain. But it is also pointed out by many researchers, with its *Sillon industriel*, 'Especially in the Haine, Sambre and Meuse valleys, between the Borinage and Liège, (...) there was a huge industrial development based on coal-mining and iron-making...'. Philippe Raxhon wrote about the period after 1830: "It was not propaganda but a reality the Walloon regions were becoming the second industrial power all over the world after Britain." "The sole industrial centre outside the collieries and blast furnaces of Walloon was the old cloth making town of Ghent." Michel De Coster, Professor at the Université de Liège wrote also: "The historians and the economists say that Belgium was the second industrial power of the world, in proportion to its population and its territory (...) But this rank is the one of Wallonia where the coal-mines, the blast furnaces, the iron and zinc factories, the wool industry, the glass industry, the weapons industry... were concentrated."

Demographic Effects

Wallonia was also the birthplace of a strong Socialist party and strong trade-unions in a particular sociological landscape. At the left, the *Sillon industriel*, which runs from Mons in the west, to Verviers in the east (except part of North Flanders, in another period of the industrial revolution, after 1920). Even if Belgium is the second industrial country after Britain, the effect of the industrial revolution there was very different. In 'Breaking stereotypes', Muriel Neven and Isabelle Devious say:

The industrial revolution changed a mainly rural society into an urban one, but with a strong contrast between northern and southern Belgium. During the Middle Ages and the Early Modern Period, Flanders was characterised by the presence of large urban centres (...) at the beginning of the nineteenth century this region (Flanders), with an urbanisation degree of more than 30 per cent, remained one of the most urbanised in the world. By comparison, this proportion reached only 17 per cent in Wallonia, barely 10 per cent in most West European countries, 16 per cent in France and 25 per cent in Britain. Nineteenth century industrialisation did not affect the traditional urban infrastructure, except in Ghent (...) Also, in Wallonia the traditional urban network was largely unaffected by the industrialisation process, even though the proportion of city-dwellers rose from 17 to 45 per cent between 1831 and 1910. Especially in the Haine, Sambre and Meuse valleys, between the Borinage and Liège, where there was a huge industrial development based on coal-mining and iron-making, urbanisation was fast. During these eighty years the number of municipalities with more than 5,000 inhabitants increased from only 21 to more than one hundred, concentrating nearly half of the Walloon population in this region. Nevertheless, industrialisation remained quite traditional in the sense that it did not lead to the growth of modern and large urban centres, but to a conurbation of industrial villages and towns developed around a coal-mine or a factory. Communication routes between these small centres only became populated later and created a much less dense urban morphology than, for instance, the area around Liège where the old town was there to direct migratory flows.

France

The industrial revolution in France followed a particular course as it did not correspond to the main model followed by other countries. Notably, most French historians argue France did not go through a clear *take-off*. Instead, France's economic growth and industrialisation process was slow and steady through the 18th and 19th centuries. However, some stages were identified by Maurice Lévy-Leboyer:

- French Revolution and Napoleonic wars (1789–1815),

- industrialisation, along with Britain (1815–1860),

- economic slowdown (1860–1905),

- renewal of the growth after 1905.

Germany

Based on its leadership in chemical research in the universities and industrial laboratories, Germany, which was unified in 1871, became dominant in the world's chemical industry in the late 19th century. At first the production of dyes based on aniline was critical.

Germany's political disunity—with three dozen states—and a pervasive conservatism made it difficult to build railways in the 1830s. However, by the 1840s, trunk lines linked the major cities; each German state was responsible for the lines within its own borders. Lacking a technological base at first, the Germans imported their engineering and hardware from Britain, but quickly learned the skills needed to operate and expand the railways. In many cities, the new railway shops were the centres of technological awareness and training, so that by 1850, Germany was self-sufficient in meeting the demands of railroad construction, and the railways were a major impetus for the growth of the new steel industry. Observers found that even as late as 1890, their engineering was inferior to Britain's. However, German unification in 1870 stimulated consolidation, nationalisation into state-owned companies, and further rapid growth. Unlike the situation in France, the goal was support of industrialisation, and so heavy lines crisscrossed the Ruhr and other industrial districts, and provided good connections to the major ports of Hamburg and Bremen. By 1880, Germany had 9,400 locomotives pulling 43,000 passengers and 30,000 tons of freight, and pulled ahead of France.

Sweden

During the period 1790–1815 Sweden experienced two parallel economic movements: an *agricultural revolution* with larger agricultural estates, new crops and farming tools and a commercialisation of farming, and a *protoindustrialisation*, with small industries being established in the countryside and with workers switching between agricultural work in summer and industrial production in winter. This led to economic growth benefiting large sections of the population and leading up to a *consumption revolution* starting in the 1820s.

During 1815–1850 the protoindustries developed into more specialised and larger industries. This period witnessed increasing regional specialisation with mining in Bergslagen, textile mills in Sjuhäradsbygden and forestry in Norrland. Several important institutional changes took place in this period, such as free and mandatory schooling introduced 1842 (as first country in the world), the abolition of the national monopoly on trade in handicrafts in 1846, and a stock company law in 1848.

During 1850–1890, Sweden experienced a veritable explosion in export, dominated by crops, wood and steel. Sweden abolished most tariffs and other barriers to free trade in the 1850s and joined the gold standard in 1873.

During 1890–1930, Sweden experienced the second industrial revolution. New industries developed with their focus on the domestic market: mechanical engineering, power utilities, papermaking and textile.

United States

During the late 18th an early 19th centuries when the UK and parts of Western Europe began to industrialise, the US was primarily an agricultural and natural resource producing and processing economy. The building of roads and canals, the introduction of steamboats and the building of railroads were important for handling agricultural and natural resource products in the large and sparsely populated country of the period.

Important American technological contributions during the period of the Industrial Revolution were the cotton gin and the development of a system for making interchangeable parts, the latter aided by the development of the milling machine in the US. The development of machine tools and the system of interchangeable parts were the basis for the rise of the US as the world's leading industrial nation in the late 19th century.

Oliver Evans invented an automated flour mill in the mid 1780s that used control mechanisms and conveyors so that no labour was needed from the time grain was loaded into the elevator buckets until flour was discharged into a wagon. This is considered to be the first modern materials handling system an important advance in the progress toward mass production.

The United States originally used horse-powered machinery for small scale applications such as grain milling, but eventually switched to water power after textile factories began being built in the 1790s. As a result, industrialisation was concentrated in New England and the Northeastern United States, which has fast-moving rivers. The newer water-powered production lines proved more economical than horse-drawn production. In the late 19th century steam-powered manufacturing overtook water-powered manufacturing, allowing the industry to spread to the Midwest.

Thomas Somers and the Cabot Brothers founded the Beverly Cotton Manufactory in 1787, the first cotton mill in America, the largest cotton mill of its era, and a significant milestone in the research and development of cotton mills in the future. This mill was designed to use horse power, but the operators quickly learned that the horse-drawn platform was economically unstable, and had economic losses for years. Despite the losses, the Manufactory served as a playground of innovation, both in turning a large amount of cotton, but also developing the water-powered milling structure used in Slater's Mill.

In 1793, Samuel Slater (1768–1835) founded the Slater Mill at Pawtucket, Rhode Island. He had learned of the new textile technologies as a boy apprentice in Derbyshire, England, and defied laws against the emigration of skilled workers by leaving for New York in 1789, hoping to make money with his knowledge. After founding Slater's Mill, he went on to own 13 textile mills. Daniel Day established a wool carding mill in the Blackstone Valley at Uxbridge, Massachusetts in 1809, the third woollen mill established in the US (The first was in Hartford, Connecticut, and the second at Watertown, Massachusetts.) The John H. Chafee Blackstone River Valley National Heritage Corridor retraces the history of "America's Hardest-Working River', the Blackstone. The Blackstone River and its tributaries, which cover more than 45 miles (72 km) from Worcester, Massachusetts to Providence, Rhode Island, was the birthplace of America's Industrial Revolution. At its peak over 1100 mills operated in this valley, including Slater's mill, and with it the earliest beginnings of America's Industrial and Technological Development.

Merchant Francis Cabot Lowell from Newburyport, Massachusetts memorised the design of textile machines on his tour of British factories in 1810. Realising that the War of 1812 had ruined his import business but that a demand for domestic finished cloth was emerging in America, on his return to the United States, he set up the Boston Manufacturing Company. Lowell and his partners built America's second cotton-to-cloth textile mill at Waltham, Massachusetts, second to the Beverly Cotton Manufactory. After his death in 1817, his associates built America's first planned factory town, which they named after him. This enterprise was capitalised in a public stock offering, one of the first uses of it in the United States. Lowell, Massachusetts, using 5.6 miles (9.0 km) of canals and 10,000 horsepower delivered by the Merrimack River, is considered by some as a major contributor to the success of the American Industrial Revolution. The short-lived utopia-like Waltham-Lowell system was formed, as a direct response to the poor working conditions in Britain. However, by 1850, especially following the Irish Potato Famine, the system had been replaced by poor immigrant labour.

The industrialisation of the watch industry started 1854 also in Waltham, Massachusetts, at the Waltham Watch Company, with the development of machine tools, gauges and assembling methods adapted to the micro precision required for watches.

Japan

The industrial revolution began about 1870 as Meiji period leaders decided to catch up with the West. The government built railroads, improved roads, and inaugurated a land reform programme to prepare the country for further development. It inaugurated a new Western-based education system for all young people, sent thousands of students to the United States and Europe, and hired more than 3,000 Westerners to teach modern science, mathematics, technology, and foreign languages in Japan (O-yatoi gaikokujin).

In 1871, a group of Japanese politicians known as the Iwakura Mission toured Europe and the United States to learn western ways. The result was a deliberate state-led industrialisation policy to enable Japan to quickly catch up. The Bank of Japan, founded in 1882, used taxes to fund model steel and textile factories. Education was expanded and Japanese students were sent to study in the west.

Modern industry first appeared in textiles, including cotton and especially silk, which was based in home workshops in rural areas.

Second Industrial Revolution

Steel is often cited as the first of several new areas for industrial mass-production, which are said to characterise a "Second Industrial Revolution", beginning around 1850, although a method for mass manufacture of steel was not invented until the 1860s, when Sir Henry Bessemer invented a new furnace which could convert molten pig iron into steel in large quantities. However, it only became widely available in the 1870s after the process was modified to produce more uniform quality. Bessemer steel was being displaced by the open hearth furnace near the end of the 19th century.

This Second Industrial Revolution gradually grew to include chemicals, mainly the chemical industries, petroleum (refining and distribution), and, in the 20th century, the automotive industry, and was marked by a transition of technological leadership from Britain to the United States and Germany.

The increasing availability of economical petroleum products also reduced the importance of coal and further widened the potential for industrialisation.

A new revolution began with electricity and electrification in the electrical industries. The introduction of hydroelectric power generation in the Alps enabled the rapid industrialisation of coal-deprived northern Italy, beginning in the 1890s.

By the 1890s, industrialisation in these areas had created the first giant industrial corporations with burgeoning global interests, as companies like U.S. Steel, General Electric, Standard Oil and Bayer AG joined the railroad and ship companies on the world's stock markets.

Opposition from Romanticism

During the Industrial Revolution an intellectual and artistic hostility towards the new industrialisation developed, associated with the Romantic movement. Romanticism revered the traditionalism of rural life and recoiled against the upheavals caused by industrialization, urbanization and the wretchedness of the working classes. Its major exponents in English included the artist and poet William Blake and poets William Wordsworth, Samuel Taylor Coleridge, John Keats, Lord Byron and Percy Bysshe Shelley. The movement stressed the importance of "nature" in art and language, in contrast to "monstrous" machines and factories; the "Dark satanic mills" of Blake's poem "And did those feet in ancient time". Mary Shelley's novel *Frankenstein* reflected concerns that scientific progress might be two-edged. French Romanticisim likewise was highly critical of industry.

Causes

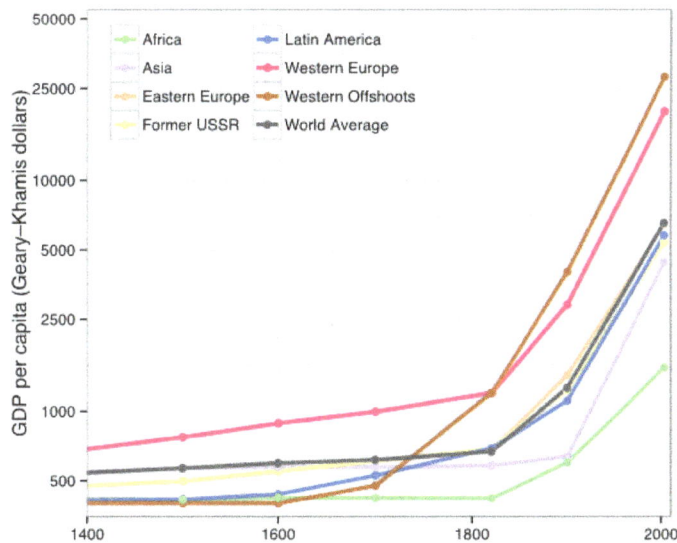

Regional GDP per capita changed very little for most of human history before the Industrial Revolution.

The causes of the Industrial Revolution were complicated and remain a topic for debate, with some historians believing the Revolution was an outgrowth of social and institutional changes brought by the end of feudalism in Britain after the English Civil War in the 17th century. As national border controls became more effective, the spread of disease was lessened, thereby preventing the epidemics common in previous times. The percentage of children who lived past infancy rose significantly, leading to a larger workforce. The Enclosure movement and the British Agricultural Revolution made food production more efficient and less labour-intensive, forcing the surplus population who could no longer find employment in agriculture into cottage industry, for example weaving, and in the longer term into the cities and the newly developed factories. The colonial expansion of the 17th century with the accompanying development of international trade, creation of financial markets and accumulation of capital are also cited as factors, as is the scientific revolution of the 17th century.

Until the 1980s, it was universally believed by academic historians that technological innovation was the heart of the Industrial Revolution and the key enabling technology was the invention and improvement of the steam engine. However, recent research into the Marketing Era has challenged the traditional, supply-oriented interpretation of the Industrial Revolution.

Lewis Mumford has proposed that the Industrial Revolution had its origins in the Early Middle Ages, much earlier than most estimates. He explains that the model for standardised mass production was the printing press and that "the archetypal model for the industrial era was the clock". He also cites the monastic emphasis on order and time-keeping, as well as the fact that medieval cities had at their centre a church with bell ringing at regular intervals as being necessary precursors to a greater synchronisation necessary for later, more physical, manifestations such as the steam engine.

The presence of a large domestic market should also be considered an important driver of the Industrial Revolution, particularly explaining why it occurred in Britain. In other nations, such as France, markets were split up by local regions, which often imposed tolls and tariffs on goods traded among them. Internal tariffs were abolished by Henry VIII of England, they survived in Russia till 1753, 1789 in France and 1839 in Spain.

Governments' grant of limited monopolies to inventors under a developing patent system (the Statute of Monopolies in 1623) is considered an influential factor. The effects of patents, both good and ill, on the development of industrialisation are clearly illustrated in the history of the steam engine, the key enabling technology. In return for publicly revealing the workings of an invention the patent system rewarded inventors such as James Watt by allowing them to monopolise the production of the first steam engines, thereby rewarding inventors and increasing the pace of technological development. However, monopolies bring with them their own inefficiencies which may counterbalance, or even overbalance, the beneficial effects of publicising ingenuity and rewarding inventors. Watt's monopoly may have prevented other inventors, such as Richard Trevithick, William Murdoch or Jonathan Hornblower, from introducing improved steam engines, thereby retarding the industrial revolution by about 16 years.

Causes in Europe

One question of active interest to historians is why the Industrial Revolution occurred in Europe and not in other parts of the world in the 18th century, particularly China, India, and the Middle East, or at other times like in Classical Antiquity or the Middle Ages. Numerous factors have been suggested, including education, technological changes, "modern" government, "modern" work attitudes, ecology, and culture. However, most historians contest the assertion that Europe and China were roughly equal because modern estimates of per capita income on Western Europe in the late 18th century are of roughly 1,500 dollars in purchasing power parity (and Britain had a per capita income of nearly 2,000 dollars) whereas China, by comparison, had only 450 dollars.

Some historians such as David Landes and Max Weber credit the different belief systems in Asia and Europe with dictating where the revolution occurred. The religion and beliefs of Europe were largely products of Judaeo-Christianity and Greek thought. Conversely, Chinese society was founded on men like Confucius, Mencius, Han Feizi (Legalism), Lao Tzu (Taoism), and Buddha (Buddhism), resulting in very different worldviews. Other factors include the considerable distance of China's coal deposits, though large, from its cities as well as the then unnavigable Yellow River that connects these deposits to the sea.

Regarding India, the Marxist historian Rajani Palme Dutt said: "The capital to finance the Industrial Revolution in India instead went into financing the Industrial Revolution in Britain." In contrast to China, India was split up into many competing kingdoms, with the three major ones being the Marathas, Sikhs and the Mughals. In addition, the economy was highly dependent on two sectors—agriculture of subsistence and cotton, and there appears to have been little technical innovation. It is believed that the vast amounts of wealth were largely stored away in palace treasuries by totalitarian monarchs prior to the British take over.

Causes in Britain

Relative Share of World Manufacturing Output, 1750-1900

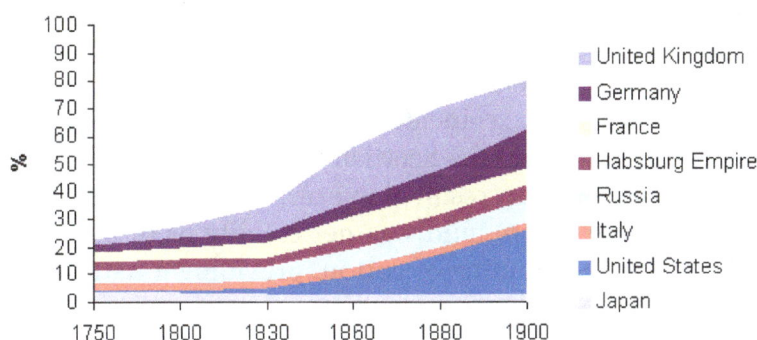

As the Industrial Revolution developed British manufactured output surged ahead of other economies. After the Industrial Revolution, it was overtaken later by the United States.

Great Britain provided the legal and cultural foundations that enabled entrepreneurs to pioneer the industrial revolution. Key factors fostering this environment were: (1) The period of peace and stability which followed the unification of England and Scotland; (2) no trade barriers between England and Scotland; (3) the rule of law (respecting the sanctity of contracts); (4) a straightforward legal system which allowed the formation of joint-stock companies (corporations); and (5) a free market (capitalism).

Geographical and natural resource advantages of Great Britain were the fact that it had extensive coast lines and many navigable rivers in an age where water was the easiest means of transportation and having the highest quality coal in Europe.

There were two main values that really drove the Industrial Revolution in Britain. These values were self-interest and an entrepreneurial spirit. Because of these interests, many industrial advances were made that resulted in a huge increase in personal wealth. These advancements also greatly benefitted the British society as a whole. Countries around the world started to recognise the changes and advancements in Britain and use them as an example to begin their own Industrial Revolutions.

The debate about the start of the Industrial Revolution also concerns the massive lead that Great Britain had over other countries. Some have stressed the importance of natural or financial resources that Britain received from its many overseas colonies or that profits from the British slave trade between Africa and the Caribbean helped fuel industrial investment. However, it has been pointed out that slave trade and West Indian plantations provided only 5% of the British national income during the years of the Industrial Revolution. Even though slavery accounted for so little, Caribbean-based demand accounted for 12% of Britain's industrial output.

Instead, the greater liberalisation of trade from a large merchant base may have allowed Britain to produce and use emerging scientific and technological developments more effectively than countries with stronger monarchies, particularly China and Russia. Britain emerged from the Napoleonic Wars as the only European nation not ravaged by financial plunder and economic collapse, and having the only merchant fleet of any useful size (European merchant fleets were destroyed during the war by the Royal Navy). Britain's extensive exporting cottage industries also ensured markets were already available for many early forms of manufactured goods. The conflict resulted in most British warfare being conducted overseas, reducing the devastating effects of territorial conquest that affected much of Europe. This was further aided by Britain's geographical position—an island separated from the rest of mainland Europe.

Another theory is that Britain was able to succeed in the Industrial Revolution due to the availability of key resources it possessed. It had a dense population for its small geographical size. Enclosure of common land and the related agricultural revolution made a supply of this labour readily available. There was also a local coincidence of natural resources in the North of England, the English Midlands, South Wales and the Scottish Lowlands. Local supplies of coal, iron, lead, copper, tin, limestone and water power, resulted in excellent conditions for the development and expansion of industry. Also, the damp, mild weather conditions of the North West of England provided ideal conditions for the spinning of cotton, providing a natural starting point for the birth of the textiles industry.

The stable political situation in Britain from around 1688, and British society's greater receptiveness to change (compared with other European countries) can also be said to be factors favouring the Industrial Revolution. Peasant resistance to industrialisation was largely eliminated by the Enclosure movement, and the landed upper classes developed commercial interests that made them pioneers in removing obstacles to the growth of capitalism. (This point is also made in Hilaire Belloc's The Servile State.)

Britain's population grew 280% 1550–1820, while the rest of Western Europe grew 50–80%. Seventy percent of European urbanisation happened in Britain 1750–1800. By 1800, only the Netherlands was more urbanised than Britain. This was only possible because coal, coke, imported cotton, brick and slate had replaced wood, charcoal, flax, peat and thatch. The latter compete with land grown to feed people while mined materials do not. Yet more land would be freed when chemical fertilisers replaced manure and horse's work was mechanised. A workhorse needs 3 to 5 acres (1.21 to 2.02 ha) for fodder while even early steam engines produced four times more mechanical energy.

In 1700, 5/6 of coal mined worldwide was in Britain, while the Netherlands had none; so despite having Europe's best transport, most urbanised, well paid, literate people and lowest taxes, it failed to industrialise. In the 18th century, it was the only European country whose cities and population shrank. Without coal, Britain would have run out of suitable river sites for mills by the 1830s.

Transfer of Knowledge

Knowledge of innovation was spread by several means. Workers who were trained in the technique might move to another employer or might be poached. A common method was for someone to make a study tour, gathering information where he could. During the whole of the Industrial Revolution and for the century before, all European countries and America engaged in study-touring; some nations, like Sweden and France, even trained civil servants or technicians to undertake it as a matter of state policy. In other countries, notably Britain and America, this practice was carried out by individual manufacturers eager to improve their own methods. Study tours were common then, as now, as was the keeping of travel diaries. Records made by industrialists and technicians of the period are an incomparable source of information about their methods.

Another means for the spread of innovation was by the network of informal philosophical societies, like the Lunar Society of Birmingham, in which members met to discuss 'natural philosophy' (*i.e.* science) and often its application to manufacturing. The Lunar Society flourished from 1765 to 1809, and it has been said of them, "They were, if you like, the revolutionary committee of that most far reaching of all the eighteenth century revolutions, the Industrial Revolution". Other such societies published volumes of proceedings and transactions. For example, the London-based Royal Society of Arts published an illustrated volume of new inventions, as well as papers about them in its annual *Transactions*.

There were publications describing technology. Encyclopaedias such as Harris's *Lexicon Technicum* (1704) and Abraham Rees's *Cyclopaedia* (1802–1819) contain much of value. *Cyclopaedia* contains an enormous amount of information about the science and technology of the first half of the Industrial Revolution, very well illustrated by fine engravings. Foreign printed sources such as the *Descriptions des Arts et Métiers* and Diderot's *Encyclopédie* explained foreign methods with fine engraved plates.

Periodical publications about manufacturing and technology began to appear in the last decade of the 18th century, and many regularly included notice of the latest patents. Foreign periodicals, such as the *Annales des Mines*, published accounts of travels made by French engineers who observed British methods on study tours.

Protestant Work Ethic

Another theory is that the British advance was due to the presence of an entrepreneurial class which believed in progress, technology and hard work. The existence of this class is often linked to the Protestant work ethic and the particular status of the Baptists and the dissenting Protestant sects, such as the Quakers and Presbyterians that had flourished with the English Civil War. Reinforcement of confidence in the rule of law, which followed establishment of the prototype of constitutional monarchy in Britain in the Glorious Revolution of 1688, and the emergence of a stable financial market there based on the management of the national debt by the Bank of England, contributed to the capacity for, and interest in, private financial investment in industrial ventures.

Dissenters found themselves barred or discouraged from almost all public offices, as well as education at England's only two universities at the time (although dissenters were still free to study at Scotland's four universities). When the restoration of the monarchy took place and membership in the official Anglican Church became mandatory due to the Test Act, they thereupon became active in banking, manufacturing and education. The Unitarians, in particular, were very involved in education, by running Dissenting Academies, where, in contrast to the universities of Oxford and Cambridge and schools such as Eton and Harrow, much attention was given to mathematics and the sciences—areas of scholarship vital to the development of manufacturing technologies.

Historians sometimes consider this social factor to be extremely important, along with the nature of the national economies involved. While members of these sects were excluded from certain circles of the government, they were considered fellow Protestants, to a limited extent, by many in the middle class, such as traditional financiers or other businessmen. Given this relative tolerance and the supply of capital, the natural outlet for the more enterprising members of these sects would be to seek new opportunities in the technologies created in the wake of the scientific revolution of the 17th century.

References

- Lucas, Robert E., Jr. (2002). Lectures on Economic Growth. Cambridge: Harvard University Press. pp. 109–10. ISBN 978-0-674-01601-9.

- Taylor, George Rogers. The Transportation Revolution, 1815–1860. ISBN 978-0-87332-101-3. No name is given to the transition years. The Transportation Revolution began with improved roads in the late 18th century.

- Roe, Joseph Wickham (1916), English and American Tool Builders, New Haven, Connecticut: Yale University Press, LCCN 16011753. Reprinted by McGraw-Hill, New York and London, 1926 (LCCN 27-24075); and by Lindsay Publications, Inc., Bradley, Illinois, (ISBN 978-0-917914-73-7).

- Crouzet, François (1996). "France". In Teich, Mikuláš; Porter, Roy. The industrial revolution in national context:Europe and the USA. Cambridge University Press. p. 45. ISBN 978-0-521-40940-7. LCCN 95025377.

- BLANQUI Jérôme-Adolphe, Histoire de l'économie politique en Europe depuis les anciens jusqu'à nos jours, 1837, ISBN 978-0-543-94762-8

- Rosen, William (2012). The Most Powerful Idea in the World: A Story of Steam, Industry and Invention. University Of Chicago Press. p. 149. ISBN 978-0-226-72634-2.

Permissions

All chapters in this book are published with permission under the Creative Commons Attribution Share Alike License or equivalent. Every chapter published in this book has been scrutinized by our experts. Their significance has been extensively debated. The topics covered herein carry significant information for a comprehensive understanding. They may even be implemented as practical applications or may be referred to as a beginning point for further studies.

We would like to thank the editorial team for lending their expertise to make the book truly unique. They have played a crucial role in the development of this book. Without their invaluable contributions this book wouldn't have been possible. They have made vital efforts to compile up to date information on the varied aspects of this subject to make this book a valuable addition to the collection of many professionals and students.

This book was conceptualized with the vision of imparting up-to-date and integrated information in this field. To ensure the same, a matchless editorial board was set up. Every individual on the board went through rigorous rounds of assessment to prove their worth. After which they invested a large part of their time researching and compiling the most relevant data for our readers.

The editorial board has been involved in producing this book since its inception. They have spent rigorous hours researching and exploring the diverse topics which have resulted in the successful publishing of this book. They have passed on their knowledge of decades through this book. To expedite this challenging task, the publisher supported the team at every step. A small team of assistant editors was also appointed to further simplify the editing procedure and attain best results for the readers.

Apart from the editorial board, the designing team has also invested a significant amount of their time in understanding the subject and creating the most relevant covers. They scrutinized every image to scout for the most suitable representation of the subject and create an appropriate cover for the book.

The publishing team has been an ardent support to the editorial, designing and production team. Their endless efforts to recruit the best for this project, has resulted in the accomplishment of this book. They are a veteran in the field of academics and their pool of knowledge is as vast as their experience in printing. Their expertise and guidance has proved useful at every step. Their uncompromising quality standards have made this book an exceptional effort. Their encouragement from time to time has been an inspiration for everyone.

The publisher and the editorial board hope that this book will prove to be a valuable piece of knowledge for students, practitioners and scholars across the globe.

Index

www.ingramcontent.com/pod-product-compliance
Lightning Source LLC
Chambersburg PA
CBHW061317190326
41458CB00011B/3827